T0181994

Wireless Networks

Series editor

Xuemin (Sherman) Shen
University of Waterloo, Waterloo, Ontario, Canada

More information about this series at http://www.springer.com/series/14180

Mojtaba Vaezi · Ying Zhang

Cloud Mobile Networks

From RAN to EPC

 Springer

Mojtaba Vaezi
Princeton University
Princeton, NJ
USA

Ying Zhang
Hewlett Packard Labs
Fremont, CA
USA

ISSN 2366-1186
Wireless Networks
ISBN 978-3-319-85407-6
DOI 10.1007/978-3-319-54496-0

ISSN 2366-1445   (electronic)

ISBN 978-3-319-54496-0   (eBook)

Printed on acid-free paper

This Springer imprint is published by Springer Nature
The registered company is Springer International Publishing AG
The registered company address is: Gewerbestrasse 11, 6330 Cham, Switzerland

Preface

In view of cost-effective implementations, scalable infrastructure and elastic capacity on demand, *virtualization* and *cloud computing* are now becoming the cornerstones of any successful IT strategy. Emerging cloud computing technologies, such as the *software-defined networking* and *Internet of things*, have been thoroughly investigated for data computing networks, while less attention has been paid to radio access network virtualization, be it hardware or software elements.

Today, the focus of research in wireless and cellular networks has shifted to *virtualization* and *cloud* technologies, so that incorporation of cloud technologies, network functions virtualization, and software-defined networking is essential part in the development process of 5G cellular communications system, expected to be commercialized by 2020. These technologies are expected to affect different parts of cellular networks including the core network and radio access network (RAN).

Cloud RAN has emerged as a revolutionary approach to implementation, management, and performance improvement of next-generation cellular networks. Combined with other technologies, such as small cells, it provides a promising direction for the zettabyte Internet era. The virtualization of RAN elements is stressing the wireless networks and protocols, especially when the large-scale cooperative signal processing and networking, including signal processing in the physical layer, scheduling and resources allocation in the medium access control layer, and radio resources managements in the network layer, are centralized and cloud computed.

The main motivation for offering this book stems from the observation that, at present there is no comprehensive source of information about cloud RAN and its interplay with other emerging technologies for network automation, such as the software-defined networking, network functions virtualization, and wireless virtualization. In addition to providing the latest advances in this area, we also include research potentials and market trend in this field. We believe that it is valuable to bring basic concepts and practical implementation of several related areas together, to facilitate a better understanding of the entire area.

Princeton, USA   Mojtaba Vaezi
Fremont, USA   Ying Zhang

Contents

Abbreviations and Acronyms

3GPP	3rd Generation Partnership Project
AAS	Active Antenna System
AIS	Alarm Indication Signal
API	Application Programming Interface
ARPU	Average Revenue Per User
ARQ	Automatic Repeat reQuest
ASIC	Application Specific Integrated Circuit
AWGN	Additive White Gaussian Noise
BBU	Baseband Unit
BRAS	Broadband Remote Access Server
BS	Base Station
BSS	Business Support System
BTS	Base Transceiver Station
CAGR	Compound Annual Growth Rate
CapEx	Capital Expenditure
CB	Coordinated Beamforming
CO_2	Carbon dioxide
CoMP	Coordinated MultiPoint
CPRI	Common Public Radio Interface
CPU	Central Processing Unit
C-RAN	Cloud RAN
CS	Coordinated Scheduling
CSI	Channel State Information
DAS	Distributed Antenna Systems
DC	Data Center
DPI	Deep Packet Inspection
D-RoF	Digital Radio over Fiber
DSL	Digital Subscriber Line
EDGE	Enhanced Data rates for GSM Evolution
eNB	Evolved NodeB

eNodeB	E-UTRAN NodeB
EPC	Evolved Packet Core
ETSI	European Telecommunications Standards Institute
eUTRAN	Evolved UTRAN
EVM	Error Vector Magnitude
FDMA	Frequency Division Multiple Access
GGSN	Gateway GPRS Support Node
GPP	General Purpose Processor
GPRS	General Packet Radio Service
GSM	Global System for Mobile Communications
GTP	GPRS Tunneling Protocol
GWCN	Gateway Core Network
HARQ	Hybrid Automatic Repeat reQuest
HetNet	Heterogeneous Network
HSPA	High Speed Packet Access
HTTP	Hypertext Transfer Protocol
IaaS	Infrastructure as a Service
ICI	Inter-Cell Interference
ICT	Information and Communications Technology
IDS	Intrusion Detection Systems
IMS	IP Multimedia Subsystem
InP	Infrastructure Provider
IoT	Internet of Things
IP	Internet Protocol
I/O	Input/Output
ISP	Internet Service Provider
ISSU	In-Service Software Upgrades
IT	Information Technology
JP	Joint Processing
JT	Joint Transmission
L1	Layer 1
L1VPN	Layer 1 VPN
L2TP	Layer 2 Tunneling Protocol
L2VPN	Layer 2 VPN
L3VPN	Layer 3 VPN
LAN	Local Area Networks
LTE	Long-Term Evolution
LTE-A	LTE-Advanced
M2M	Machine-to-Machine
MAC	Media Access Control
MANO	Management and Network Orchestration
MIMO	Multiple Input Multiple Output
MIP	Mixed Integer Programming
MME	Mobility Management Entity
mmW	Millimeter Wave

MOCN	Multi-Operator Core Network
MSC	Mobile services Switching Center
MU-MIMO	Multi-User MIMO
MVNO	Mobile Virtual Network Operators
MWC	Mobile World Congress
NAT	Network Address Translation
NF	Network Functions
NFV	Network Functions Virtualization
NFVI	Network Functions Virtualization Infrastructure
NIC	Network Interface Card
NOMA	Non-orthogonal Multiple Access
NV	Network Virtualization
O&M	Operations & Maintenance
OBSAI	Open Base Station Architecture Initiative
OFDMA	Orthogonal Frequency Division Multiple Access
OpEx	Operational Expenditure
OSPF	Open Shortest Path First
OSS	Operational Support System
OTN	Optical Transmission Network
P2P	Peer-to-Peer
PaaS	Platform-as-a-Service
PDCP	Packet Data Convergence Protocol
PDH	Plesiochronous Digital Hierarchy
PE	Provider Edge
PGW	Packet Data Networks Gateway
PHY	Physical Layer
PON	Passive Optical Network
POP	Point of Presence
PSTN	Public Switched Telephone Network
QoE	Quality of Experience
QoS	Quality of Service
RAN	Radio Access Networks
RANaaS	RAN-as-a-Service
RAT	Radio Access Technologies
RFIC	Radio Frequency Integrated Circuit
RLC	Radio Link Control
RNC	Radio Network Controller
RRC	Radio Resource Control
RRH	Remote Radio Head
SaaS	Software-as-a-Service
SBT	Session Border Controllers
SC-FDMA	Single Carrier Frequency Division Multiple Access
SDH	Synchronous Digital Hierarchy
SDN	Software-Defined Networking
SGSN	Serving GPRS Support Node

SGW	Serving Gateway
SINR	Signal-to-Interference-plus-Noise Ratio
SLA	Service-Level Agreement
SON	Self-Organizing Networking
SONET	Synchronous Optical Networking
SP	Service Provider
SSL	Secure Sockets Layer
TCO	Total Cost of Ownership
TDMA	Time Division Multiple Access
TMA	Tower Mounted Amplifiers
TP	Transmission Point
TTI	Transmission Time Interval
UE	User Equipment
UTRAN	Universal Terrestrial Radio Access
vCPU	virtual CPU
vEPC	virtualized EPC
vIMS	virtualized IMS
VIM	Virtualized Infrastructure Management
VLAN	Virtual Area Network
VLR	Visitor Location Register
vNF	virtual Network Functions
vNIC	virtualized NIC
VoIP	Voice over IP
VPN	Virtual Private Network
vRAM	virtual RAM
VSN	Virtual Sharing Network
WAN	Wide Area Network
WiFi	Wireless Fidelity
WiMAX	Worldwide Interoperability for Microwave Access
WNC	Wireless Network Cloud
XaaS	X-as-a-Service

List of Figures

List of Tables

Chapter 1
Introduction

Over the past decades, the telecommunications industry has migrated from legacy telephony networks to telephony networks based on an IP backbone. IP-based networks have offered operators the opportunity to access previously untapped networks/subscribers to offer innovative products and services, and stimulate a new wave of revenue generation. The use of smart phones, tablets, and other data consuming devices, such as machine-to-machine (M2M) modules, has explosively increased during past years, and is changing our lives in ways we did not envision. Every day more people watch more video and run more data-hungry applications using such devices. New applications are being developed on a daily basis, and M2M devices are integrated into more areas of life and industry.

Mobile communications experienced a major breakthrough when for the first time total mobile data traffic topped mobile voice traffic at the end of 2009 [10, 11], resulting in a paradigm shift from low bandwidth services, such as voice and short message, to broadband data services, such as video and online gaming. Figure 1.1 [1] shows the total global data and voice traffic in mobile networks during the past 5 years. While voice traffic is almost flat, data traffic has experienced a stable exponential growth. As an example, mobile data traffic has increased nearly 60 Also, mobile data traffic in the first quarter of 2014 has exceeded total mobile data traffic in 2011 [12].

Mobile networks will face even more increase in data traffic in coming years. According to Cisco visual networking index forecast [13, 14], by the end of 2016 global yearly IP traffic will pass the zettabyte (10^{12} gigabytes) threshold, and traffic from wireless and mobile devices will surpass traffic from wired devices. In addition, it is projected that by 2018:

- Global Internet traffic will be equivalent to 64 times of that in 2005
- Data center virtualization and cloud computing will grow remarkably and nearly one-fifth (78%) of workloads will be processed by cloud data centers
- Mobile data traffic will increase 11-fold compared with that in 2013, representing a compound annual growth rate of 61% between 2013 and 2018
- Busy-hour Internet traffic will grow by a factor of 3.4 (21% faster than average Internet traffic growth).

© Springer International Publishing AG 2017
M. Vaezi and Y. Zhang, *Cloud Mobile Networks*, Wireless Networks,
DOI 10.1007/978-3-319-54496-0_1

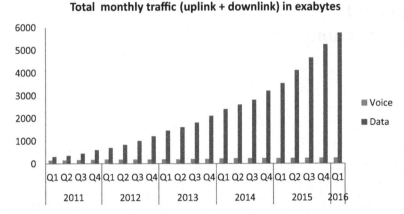

Fig. 1.1 Global quarterly data and voice traffic in mobile networks from 2011 and 2016 [1]

This data flood is caused by different factors. First of all, much more devices will access Internet and broadband services. There will be nearly 21 billion devices—including fixed and mobile personal devices and M2M connections—connected to IP networks [13]; this is 2.7 times as high as the world population, up from 1.7 networked devices per capita in 2013. Being integrated into more areas of life and industry, M2M devices annually will produce 84% more data, between 2013 and 2018. Furthermore, devices, particularly smart phones, will be more powerful and services will be more diverse and bandwidth-hungry.

To accommodate the explosively increasing data traffic, many operators see 4G technologies such as LTE as a way. However, they cannot simply upgrade their network from 2G/3G to 4G overnight because their subscribers may not upgrade their devices for several years. This implies that operators will need to support multiple technologies, at least over a period of time, so as to support a mix of subscriber technologies. More importantly, they need to make a huge investment to upgrade network infrastructure. As a result, operators are not seeing proportional revenue growth with the data traffic, and most of them facing flat to declining revenues. Besides, admittedly, even the current 4G networks are not able to address the onslaught of users demand for mobile data, and network capacity expansion is necessary.

Capacity expansion is one of the most significant technological challenges operators face today. Despite implementation of advanced capacity-improving techniques such as multi-antenna systems and increasing use of Wi-Fi offload to increase network capacity, operators can not keep up with exploding capacity needs of customers. There are multiple other techniques, such as adding spectrum and using smaller cells, that enable operators to expand their network capacity. These solutions are however, costly, difficult, and/or take time. Moreover, although bringing up considerable performance gains, they are unlikely to be able to carry the exponentially growing wireless data traffic. Besides, when a large group of smart phone users gather at a concert or an arena, for example, heavy usage can exhaust local capacity. Under such circumstances, traditional radio access network (RAN) architecture is facing more and more challenges. In addition to the unforeseen capacity challenges, requirement for a dedicated equipment room with supporting facilities for each base station increases network deployment difficulty because cell site costs and space often limit appropriate locations. Furthermore, power consumption and operating costs go up.

The current RAN architecture is not capable of addressing the explosive need of data demand; it has run its course. This is partly because it does not efficiently utilize resources, such as spectrum and processing power, as, with dedicated resources, certain cells may experience congestion while others are underutilized. Note that resource allocation in each base station is based on busy-hour traffic (the peak level of traffic), whereas busy-hour traffic can be much higher than average traffic in many sites. This is expected to become even worse over time, since busy-hour Internet traffic is growing faster than average Internet traffic [13, 14]. To address these issues and meet the growing demand, disruptive solutions are required. Possibly, the gap will have to be filled by a new generation of networks such as cloud-based networks, which offer the benefits of cloud computing in RAN [15, 16].

1.1 Challenges of Today's RAN

Conventional RANs are based on industry-specific hardware owned and operated by mobile operators. Upgrading technology, reducing ownership costs, and decreasing carbon dioxide emission at cellular sites are today's technological, economical, and ecological challenges for operators around the globe. As elaborated in the following sections, these issues stem from the architecture of legacy RAN which is based on dedicated resources for each cell site.

1.1.1 Cost of Ownership

Mobile operators are facing pressure on ever-increasing cost of ownership with much less growth of income, in the era of mobile Internet. Total cost of ownership (TCO) is an accounting term that attempts to quantify the direct and indirect financial impact of a product or system on its life cycle. TCO is designed to look at an investment from a strategic point of view; it can be divided into two main streams: capital expenditure (CAPEX) and operation expenditure (OPEX). CAPEX is generally associated with buying fixed assets or extending the life of an asset and OPEX includes the costs related to the operation, maintenance, and administration of an investment.

Figure 1.2 illustrates the TCO for cellular systems. In a mobile network, CAPEX comprises RAN infrastructure (including base station, tower, antennas and feeder, and microwave) and supplementary equipment, such as power and air-conditioning, backhaul transmission, and core network, whereas OPEX covers maintenance, cost of power, rent, and spectrum licence, field-services, planning and optimization.

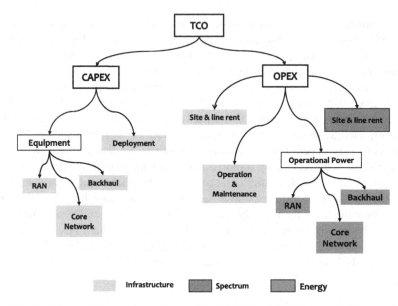

Fig. 1.2 Breakdown of the total cost of ownership (CAPEX and OPEX) for cellular networks [2]

Roughly speaking, based on several independent estimations or actual cost calcu-
lations, CAPEX comprises one-third of TCO and the other two-third belongs to
OPEX, when considering a period of 7 years for operation. This, however, varies
from network to network.[1]

Varying based on countries, it is known that 70–90% of CAPEX of a mobile
operator is spent on the RAN [8, 18, 19]. This means that most of the CAPEX
is related to building out cell sites and purchasing the equipment for the RAN.
Then, when we break it down, it appears that the cost of wireless and transmission
equipment, on average, makes nearly 40% of total CAPEX (and more than half of
the RAN CAPEX) [8, 19, 20], and other costs including site acquisition, civil works
and construction, and installation, on average, account for about another 40% of the
total CAPEX of a mobile network. The remaining 20% is spent on other parts of the
network, such as core network and backhaul.

Considering the big share of RAN in total CAPEX and the fact that more than
half of the RAN CAPEX is not spent on wireless equipment, it makes much sense
to lower the expenditure on site construction, installation, and deployment. Hence,
the focus of the mobile operators is to reduce the cost of auxiliary installations and
engineering construction, so as to lower the CAPEX of their mobile networks.

---

[1]For example, the total U.S. LTE infrastructure OPEX was anticipated to be 57.4 billion, while
CAPEX was projected to be only 37.7 billion between 2012 and 2017 [17].

1.1.2 Capacity Expansion

The unprecedented growth of mobile data will continue to gain momentum in the coming years. Adding capacity seems like the obvious answer to meet this challenge; this is not simple, though. Evolution of wireless technologies has provided higher spectral efficiency, but it has not been enough and it is difficult to meet this demand by adding spectrum. Increasing the cell deployment density of the network is another possible solution, and cell splitting is one of the main trends in RAN that enables network densification.

Vendors have introduced a wide variety of radio access nodes, such as *small cells* and *distributed antenna systems*, and operators are increasingly augmenting traditional macro expansion with network offloading solutions. Small cells, used for increasing capacity and coverage and covering high-traffic public areas, are expected to account for a large proportion of the offloaded traffic [21]. In addition, by increasing the cell deployment density of the network, average distances between a user and the nearest base station decrease. Hence, the link quality improves which results in a larger capacity for the link [22]. While small cells are viewed as an offload technique in 3G networks, by introduction of *heterogeneous network* (HetNet), they are an integral part of 4G networks.

Although small cells require low power and low cost nodes by definition, today, they are required to support multiple technologies and multiple frequency bands. By increasing the implementation of small cells, it is expected to have 50 million base stations by 2015 [23, 24], and some even predict that in 10–15 years, there may be more base stations than the number of cell phone subscribers [23]. Hence, considering cell site costs and space requirements, it is clear that adding capacity through small cells is difficult and expensive. Small cells will be an integral part of future networks, but they are not cost-effective, environment-friendly solutions. Nor are they capable of addressing the long-term mobile data capacity requirements.

1.1.3 Energy Consumption and Carbon Emissions

Information and communications technology (ICT) is one of the major components of the world energy consumption budget. It is estimated that the global ICT now accounts for about 10% of the world energy consumption [25–28]. The electricity consumption of communication networks has been growing by a compound annual growth rate (CAGR) of 10% per year during 2007–2012, two times faster than the other sectors in ICT and more than threefold greater than the growth of worldwide electricity consumption in the same time frame [27, 29]. This is mainly because during the past few years, operators have been increasingly implementing new cell sites to offer broadband wireless services, as we mentioned previously, and it was foreseen to have 50 million base stations by 2015 [23].

With the explosive growth of mobile communications in terms of number of connected devices, and the demand for new services and ubiquitous connectivity, the energy consumption of wireless access networks is increasing significantly. In particular, power consumption rises as more base stations are deployed since it is estimated that base stations contribute to 60–80% of the total energy consumption [30]. This situation imposes a major challenge for mobile operators since a higher power consumption is directly translated to a higher operational expenditures. Carbon dioxide (CO_2) emission is another important consequence of increasing energy consumption, in addition to rising OPEX. Mobile cellular communication is thought to account for 2–3% of global CO_2 emissions [31]. In 2011, Bell Labs estimated that mobile base stations globally emit about 18 million metric tons of CO_2 per year [32]. This brings about significant environmental impact and is against the current social and political trend and commitments towards a greener communication.[2]

The above economical and ecological challenges urge mobile operators to support RAN architecture and/or deployment scenarios that can cope with the traffic and network growth in a more energy-efficient manner. *Cloud RAN*, together with *software-defined networking* and *network functions virtualization*, is among the emerging technologies starting to break the traditional cellular infrastructure in general, and the radio access network, in particular.

1.2   Cloud RAN - What Is the Big Idea?

Cloud radio access network (C-RAN) architecture is currently a hot topic in the research, industry, and standardization communities. The basic idea behind C-RAN is to change the traditional RAN architecture in a way that it can take advantage of technologies like *cloud computing*, *wireless virtualization*, and *software-defined networking*. More specifically, C-RAN is a RAN architecture in which dedicated cell site base stations are replaced with one or more remote clusters of centralized virtual base stations, each of which is able to support a great many remote radio/antenna units. C-RAN may also stand for *centralized RAN*. Centralized RAN is basically an evolution of the current distributed base stations, where the baseband unit (BBU) and remote radio head (RRH) can be spaced miles apart. Centralized RAN and cloud RAN can be considered as two sides of the same coin. Although some people may use these two terms interchangeably, there is a clear difference between them. Cloud RAN implies that the network is "virtualized" on top of being centralized, meaning that it is implemented in generic server computers (or blade servers) and base station resources can be added as per their needs, to efficiently handle the network traffic. Depending on the function splitting between BBU and RRH, cloud RAN partly or wholly centralizes the RAN functionality into a shared BBU pool or cloud which is connected to RRHs in different geographical locations.

---

[2]As an example, the UK is committed to reducing the amount of CO_2 it emits in 2050 to 20% of that seen in 1990 [31].

1.2.1 Advantages

Cloud RAN has strategic implications on operator–vendor relationship, as it allows operators to implement network upgrades more agilely and to select between vendors easily. Aside from the strategic implications, the C-RAN architecture has practical and measured benefits to the current RAN, that essentially revolved around reducing the cost of network operations. We briefly review some of them here.

Major Savings in CAPEX and OPEX

From a business perspective, C-RAN is expected to bring in significant reductions in both CAPEX and OPEX due to reduced upgrading and maintenance costs. Cost saving in CAPEX is due to the fact that single cells are not required to be dimensioned for peak-hour demands. Instead, baseband processing power can be pooled and assigned specifically where needed, implying that dimensioning can be done for a group of cells rather than a single one. This increases the processing utilization largely. Also, baseband processing can be cost effectively run on commercial servers. Further CAPEX savings can be achieved from potential technology enhancements (e.g., LTE-Advanced features) which leave further processing headroom. In addition, less costly *general-purpose processor* hardware can be used for RAN algorithms.

OPEX savings can be drawn mainly from energy savings, reduced cost of maintenance, and smaller footprint required. Generally, operations and maintenance (O&M) for distributed hardware (conventional RAN architecture) is more costly than that of centralized hardware (cloud RAN architecture). Also, with C-RAN centralized network analysis and optimization, such as centralized self-organizing networks can be naturally evolved which are able to transform network economics. Besides, due to smaller footprint required at each site, site rental and civil works costs drop.

Flexibility in Network Capacity Expansion

From the network capacity expansion point of view, C-RAN brings in significant gain and flexibility. In a heterogeneous network C-RAN, low power RRH can be deployed in the coverage area of a macro cell where a high level of coordination between the macro cell and the RRH is achievable. This can reduce interference when some LTE-advanced technologies such as coordinated multipoint (CoMP), where multiple base stations transmit and receive from a mobile device, are deployed.

Reducing Energy Consumption and CO_2 Emissions

From an ecological perspective, C-RAN architecture is preferred to the conventional RAN architecture, as it consumes less energy and is greener. This is mainly because multiple BBUs can share facilities, e.g., air-conditioning, and partly because of resource aggregation which results in an improved resource utilization, which in turn improves energy efficiency. On the same page, in the centralized architecture, if required, e.g., when traffic demand is low, BBU resources can be switched off much more easily than the conventional distributed RAN. This brings in further energy savings. It is obvious that such an architecture is more environmental friendly and reduces CO_2 emissions.

    In addition to the above-mentioned benefits, adopting C-RAN allows different levels of sharing in access network. It allows the operator to efficiently support multiple radio access technologies (RAT), network sharing (sharing base band processing, RRH, and spectrum), or outsourcing. It should be mentioned that the implementation of a centralized RAN is easier than a cloud RAN. It, however, lacks the benefits associated with virtualization and cloudification.

1.2.2 Challenges

Although the basic ideas of C-RAN are already relatively mature, we are still in its early days and much work is yet required to achieve this vision. There are many challenges such as fronthaul (between BBUs or RRHs) requirements in terms of bit rate, latency, jitter and synchronization, interface definitions between BBUs and RRHs and between BBUs, and base station virtualization technology. Besides, current general-purpose processors (GPPs) are not a practical solution for handling the datapath processing and the very high data rates required by 4G systems. Also, these GPPs are not optimal platforms for such operations in terms of power consumption.

1.3 Related Technologies

Virtualization is a key enabler of cloud computing and cloud-based infrastructures. There are also other emerging technologies, such as *network functions virtualization* (NFV) and *software defined networking* (SDN), that support cloud environments. These technologies move the networking industry from today's manual configuration to more automated and scalable solutions. They are complementary approaches that solve different subsets of network mobility problem.

    Wireless carriers are targeting the integration of NFV and SDN across multiple areas including radio access network, core network, backhaul, and operational/business support systems (OSS/BSS), caused by the promise of total cost of ownership reduction. SDN and NVF are among other initiatives to move from the traditional cellular infrastructure toward a cloud-based infrastructure where RAN, mobile core, etc. are expected to be applications that can run on general-purpose infrastructure, rather than proprietary hardware, hosted by data center operators and other third parties. We briefly explain these enabling/related technologies and their relation to C-RAN in the following sections.

1.3.1 Network Virtualization

Virtualization is a technology that enables us to go beyond the physical limitations normally associated with entire classes of hardware, from servers and storage to networks and network functions. *Network virtualization* (NV) ensures that network can integrate with and support the demands of virtualized architectures. It can create a virtual network that is completely separate from other network resources. Virtualization mechanisms are at the core of cloud technologies.

1.3.2 Network Functions Virtualization

Network functions virtualization (NFV) provides a new way to design, deploy, and manage network services. It decouples the network functions from purpose-built hardware, so they can run in software. Therefore, NFV enables the implementation of services on the general-purpose hardware, allowing operators to push new services to the network edge, i.e., to base stations. This in turn helps operators support more innovative location-based applications and reduces the backhaul traffic by shifting services away from the network core.

1.3.3 Software-Defined Networking

Software-defined networking (SDN) is a new approach to designing, building, and managing networks which enables the separation of the network's *control plane* and *data plane*, which makes it easier to optimize each plane. SDN has the potential to make significant improvements to service request response times, security, and reliability. In addition, by automating many processes that are currently done manually it could reduce costs. SDN is a natural platform for network virtualization as, in a software-defined network, network control is centralized rather than being distributed in each node [33]. Therefore, this technology can be applied to C-RAN environments to enable universal management capabilities, allowing operators to remotely manage their network.

In summary, NV, NFV, and SDN each provide a software-based approach to networking, in order to make networks more scalable and innovative. Hence, expectedly, some common beliefs guide the development of each. For example, they each aim to move functionality to software, use general-purpose hardware in lieu of purpose-built hardware, and support more efficient network services. Nevertheless, note that SDN, NV, and NFV are independent, though mutually beneficial. Finally, by applying the concepts of SDN and NFV in a C-RAN environment, most of the processing can be implemented in commodity servers rather than proprietary appliances. Hence, when combined with SDN and NFV, C-RAN provides operators with reduced equipment costs and power consumption.

1.4 Outline of Chapters

This book is divided into five chapters and provides information on the different technologies enameling cloud RAN.

Chapter 1: Introduction. In this chapter, we have introduced the challenges of conventional radio access network (RAN) and the requirements for future RAN. This was followed by a brief overview of cloud RAN and its advantages. We have also outlined the developing technology relevant to cloud RAN.

Chapter 2: Wireless Virtualization. This chapter reviews various wired network virtualization technologies as well as network functions virtualization. It then studies the state of the art in the wireless virtualization and its advantages and challenges. The last part of this chapter is devoted to the cloud computing, its service models, and its relation to virtualization.

Chapter 3: Software Defined Networking. In this chapter, we provide a review of the SDN technology and business drivers, describe the high-level SDN architecture and principles, and give three scenarios of its use cases in mobile access aggregation networks and the cloud networks. Furthermore, we provide discussions on the design implementation considerations of SDN in the mobile networks and the cloud, in comparison with traditional networks.

Chapter 4: Virtualizing the Network Services: NFV. In this chapter, we provide a survey of the existing Network Function Virtualization (NFV) technologies. We first present its motivation, use cases, and architecture. We then focus on its key use case, the service function chaining, and the techniques and algorithms.

Chapter 5: SDN/NFV Telco Case Studies. In this chapter, we review the two important case studies of SDN and NFV's usage in telecom mobile network. In particular, we discuss network virtualization's usage in packet core network and in customer premise equipment (mobile edge networks). In both case studies, we discuss the challenges and the opportunities.

Chapter 6: RAN Evolution. The main objective of this chapter to set the stage to better understand and needs for the cloud RAN architecture, in Chap. 7, and its barriers and/or competing technologies. The chapter starts with an overview of the architecture of mobile networks with an emphasis on RAN and backhaul/fronthaul solutions. It then compares the legacy and distributed base stations technologies. It also describes the current and emerging trends in wireless network densification as well as several concepts related to the cloud RAN solution, including small cell, distributed antenna systems, and mobile network sharing.

Chapter 7: Cloud RAN. In this chapter, first the cloud RAN is defined, its driving concepts are elaborated, and its vision and mission are identified. Then, different implementation scenarios for the cloud RAN are studied in detail and compared with. Next, the conclusions are drawn on the possible solution for future networks with a view on the competing solutions such as small cells and edge cloud. Finally, the challenges and research frontiers are identified and described.

Chapter 2
Virtualization and Cloud Computing

In an effort to move the networking industry from today's manual configuration to embrace automated solutions that are coordinated with the rest of the infrastructure, there have been several emerging technologies in the past few years, chief among them are *network virtualization* (NV), *network functions virtualization* (NFV), and *software-defined networking* (SDN).[1] Broadly speaking, all these three solutions are designed to make networking more automated and scalable to support virtualized and cloud environments. These technologies are software-driven schemes that promise to change service and application delivery methods, so as to increase network agility. They are different but complementary approaches (with some overlapping terminology) to provide network programmability. In other words, they solve different subsets of the macro issue of *network mobility*.

It is important to mention that, despite the fact that SDN, NV, and NFV are mutually beneficial, they are not dependent on one another. That is, NFV and NV can be implemented without an SDN being required and vice versa, but SDN makes NFV and NV more compelling and vice-versa.

In this chapter, we review the state of the art in network virtualization and investigate the challenges that must be addressed to realize a viable network virtualization environment.

2.1 What Is Virtualization?

In computing, *virtualization* is the process of abstracting computing resources such that multiple applications can share a single physical hardware. Put differently, virtualization refers to the creation of a *virtual*, rather than actual, version of a resource. The canonical example of virtualization is "server virtualization," in which certain attributes of a physical server are decoupled (abstracted) and reproduced in

---

[1]The first two technologies are covered in this chapter whereas SDN is discussed in the next chapter.

© Springer International Publishing AG 2017   11
M. Vaezi and Y. Zhang, *Cloud Mobile Networks*, Wireless Networks,
DOI 10.1007/978-3-319-54496-0_2

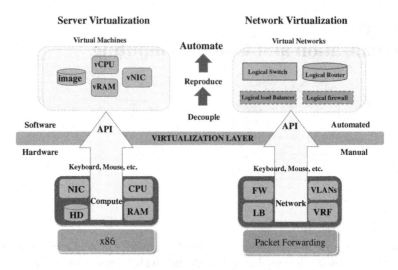

Fig. 2.1 Server virtualization: the physical hardware resources are mapped to multiple virtual machines, each with its own CPU, memory, disks, and I/O devices

a hypervisor (virtualization software) as vCPU, vRAM, vNIC, etc.; these are then assembled arbitrarily to produce a virtual server, in few seconds.

Computing resources are not the only resources that are virtualized; *storage* can be virtualized too. Through virtualization either one resource is shared among multiple users or multiple resources, e.g., storages, are aggregated and presented as one or more high capacity resource that can be used by one or multiple users. In any of those cases, the user has the illusion of sole ownership.

Besides computing and storage, a *network* may be virtualized too. That is, the notion of abstraction can be extended from computing resources and storage to the fundamental components of the networks, i.e., nodes and links. Therefore, in a broader context, virtualization refers to the creation of a *virtual* version of a resource, such as an operating system, a storage device, or network resources.

Server and desktop virtualization is a mature technology now. Virtualization software, e.g., VMware Workstation [34], maps the physical hardware resources to the virtual machines that encapsulate an operating system and its applications, as shown in Fig. 2.1. Each virtual machine fully equivalent of a standard x86 machine, as it has its own central processing unit (CPU), memory, disks, and I/O devices.

Thus far, it should be clear that with virtualization a system pretends to be more than one of the same system. In the following, we will see why this property is important and how it can make networking more programmable and agile [35].

2.2 Why Virtualization?

There are three major technical benefits of improving availability, enabling mobility, and improving utilization, but the benefits do not stop there. They extend to simplifying the IT architecture, and ultimately to accurately aligning billing with consumption.

The basic motivation for virtualization is to efficiently share resources among multiple users. This is similar to multitasking operating systems where, rather than doing one task at a time, unused computing power is used to run another task. Consider an organization that has many servers all doing single or a small cluster of related tasks. Without losing the security of isolated environments, virtualization allows these servers to be replaced by a single physical machine which hosts a number of virtual servers. Similarly, storage aggregation enhances the overall manageability of storage and provides better sharing of storage resources.

Migration is another big advantage of virtualization. It comes in handy if an upgrade is required or when the hardware is faulty because it is fairly simple to migrate a virtual machine from one physical machine to another. Therefore, increasing backup capability is another compelling reason for virtualization. If a server crashes, the data on that server can be set to be automatically transferred to another server in the network. Such a redundancy increases availability, too. What is more, a virtual machine offers a much greater degree of isolation [36].

Saving on physical machine costs, reduced energy consumption, and smaller physical space requirement are among other notable advantages of virtualization. There are also business benefits for a virtualized enterprise, including flexible sourcing, self-service consumption, and consumption-based billing. Most of those advantages are applicable to NV and/or NFV. These two paradigms, however, provide other unique advantages that will be discussed in detail, later in this chapter. Among them are, improving security [37, 38], accelerating time-to-market, and extending accessibility.

2.3 Network Virtualization

Network virtualization refers to the technology that enables partitioning or aggregating a collection of network resources and presenting them to various users in a way that each user experiences an isolated and unique view of the physical network [39–41]. The abstraction of network resources may include fundamental resources (i.e., links and nodes) or derived resources (topologies) [39]. This technology may virtualize a network device (e.g., a router or network interface card (NIC)) a link (physical channel, data path, etc.), or a network. As a result, similar to server virtualization which reproduces vCPU, vRAM, etc., network virtualization software may reproduce logical channel (L1), logical switches, logical routers (L2–L3), and

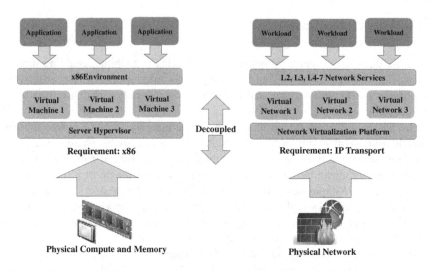

Fig. 2.2 Network virtualization versus server virtualization

more. These logical resources, along with L4–L7 services,[2] can be assembled in an arbitrary topology, presenting a complete L1–L7 virtual network topology. Thanks to network virtualization, multiple logical networks can coexist and share a physical network.

Network virtualization decouples the roles of the traditional Internet service providers (ISPs) into infrastructure providers (InPs) and service providers (SPs) [40]. InPs and SPs are two independent entities: the former manages the physical infrastructure whereas the latter creates virtual networks by aggregating resources from one or multiple InPs and offers end-to-end services. This decoupling will proliferate deployment of coexisting heterogeneous networks free of the inherent limitations of the existing Internet [40–42]. It is also a way to automate the network to improve networking administrators' responsiveness to change. Indeed, it is hard to keep up with too many requests for network configuration changes, that can take days or weeks to handle. By allowing multiple heterogeneous networks to cohabit on a single physical architecture, network virtualization increases flexibility, security, and manageability of networks (Fig. 2.2).

In view of the great degree of flexibility and manageability its offers, network virtualization has become a popular topic of interest, both in academia and industry, during recent years. However, the term network virtualization is somewhat overloaded and several definitions, from different perspectives, can be found in the literature (cf. [40, 43, 44]). The concept of multiple coexisting logical networks over a shared physical network has frequently appeared in the networking literature under several different names. Chowdhury and Boutaba [40, 41] classify them into four

---

[2]L4–L7 services also can be virtualized to produce logical load balancers or logical firewalls, for example. This is referred to as *network functions virtualization* and will be discussed in Sect. 2.4.

main categories of virtual local area networks (VLANs), virtual private networks (VPNs), overlay networks, and active and programmable networks, whereas in a recent paper [39], Wang et al. divide them into three main groups of VPNs, overlays, and virtual sharing networks (VSNs). We follow the later one due to its clearer and simpler calcification.

Broadly speaking, three types of commercial virtual networks exist which are described in the following.

2.3.1 Overlay Networks

An *overlay network*[3] is a logical network that runs independently on top a physical network (underlay). Overlay networks do not cause any changes to the underlying network. Peer-to-peer (P2P) networks, virtual private networks (VPNs), and voice over IP (VoIP) services such as Skype are examples of overlay networks [40, 41, 45]. Today, most overlay networks run on top of the public Internet, while the Internet itself began as an overlay running over the physical infrastructure of the public switched telephone network (PSTN). The Internet started by connecting a series of computers via the phone lines to share files and information between governmental offices and research agencies. Adding to the underlying voice-based telecommunications network, the Internet layer allowed data packets transmission across the public telephone system, without changing it.

P2P networks are an important class of overlay networks [46]; they use standard Internet protocols to prioritize data transmission between two or more remote computers in order to create direct connections to remote computers, for file sharing. P2P networks use the physical network's topology, but outsource data prioritization and workload to software settings and memory allocation.

Although there are various implementations of overlays at different layers of the network stack, most of them have been implemented in the application layer on top of IP, and thus, they are restricted to the inherent limitations of the existing Internet.

2.3.2 Virtual Private Networks

Many companies have offices spread across the country or around the globe, and they need to expand their private network beyond their immediate geographic area, so as to keep fast and reliable communications among their offices. Until recently, such a communication has meant the use of *leased lines* to deploy a *wide area network* (WAN). A WAN is preferred to a public network (e.g., the Internet) for its reliability, performance, and security. But, maintaining a WAN is expensive, especially when leased lines are required. What is more, leased lines are not a viable solution if

---

[3]Here, the network refers to a telecommunication or computer network.

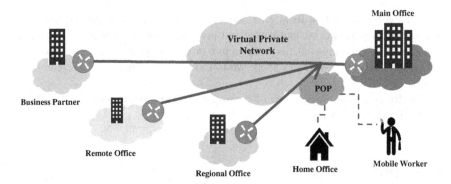

Fig. 2.3 Typical VPN topology: a private network deployed using a public network (usually the Internet) to securely connect remote sites/users together

part of the employees need to access the corporate network remotely, from home, from the road, or from other organizations. On account of ubiquitous Internet, many companies create their own virtual private networks (VPNs) to accommodate the needs of the remote (mobile) workforce and distant offices.

A virtual private network (VPN) is an assembly of two or more private networks or individual users that uses secured tunnels over a public telecommunication infrastructure, such as the Internet, for connection. A VPN is commonly used to provide distributed offices or individual users with secure access to their organization's network. This is illustrated in Fig. 2.3. It is *virtually* private as it uses *public* infrastructure to provide remote access; this access is, however, secure as if the organization uses its private (owned or leased) lines for remote connection. VPNs are meant to provide the organizations with the same capabilities as WANs but at a much lower cost. They can be remote access (connecting an individual user to a network) or site-to-site (connecting two networks together) [47].

Although VPNs use the shared public infrastructure, they maintain privacy through security procedures and *tunneling*[4] protocols such as secure socket tunneling protocol (SSTP), point-to-point tunneling protocol (PPTP), and layer two tunneling protocol (L2TP). These protocols encrypt the data and send it through a "tunnel" that cannot be entered by data that is not properly encrypted. It is worth mentioning that there is another level of encryption for the originating and receiving network addresses.

A well-designed VPN needs to incorporate security, reliability, scalability, network management, and policy management [47]. Such a network not only extends geographic connectivity, but can improve security, simplify network topology, reduce

---

[4]Tunneling is a mechanism used to send unsupported protocols across different networks; it allows for the secure movement of data from one network to another. Specifically, tunneling refers to the transmission of data intended to be used only within a private network through a public network such that the routing nodes in the public network are oblivious to the fact that the transmission is part of a private network [48].

transit time, and transportation costs for remote users, provide global networking opportunities and telecommuter support, reduce operational costs compared to traditional WANs, and improve productivity [47]. The above benefits motivate the organizations to deploy VPNs; there are other reasons, primarily for individuals, to start using a VPN [49], for example,

- One can use a VPN to connect securely to a remote network via the Internet. Most companies and nonprofit organizations, including universities, maintain VPNs so that employees can access files, applications, and other resources, from home or from the road, without compromising security.[5]
- Where online privacy is a concern, connecting to a VPN is a smart, simple security practice, when you are on a public or untrusted network, such as a Wi-Fi in a hotel or coffee shop. It helps prevent others who may be trying to capture your passwords.
- VPNs turn out to be very useful in circumventing regional restrictions (censorship) on certain websites. They can also be used for recreational purposes; for example, one can connect to a US VPN to access the US-only websites outside the US. Note that, many media websites (CNN, Fox, Netflix, etc.) may impose a geographical restriction on viewing their content and online videos.
- A VPN comes in handy to virtually "stay home away from home." Meaning that, one can virtually reside in a specific country from abroad with an IP of that country. It happens that we need to access our geographically restricted accounts, such as online banking and state websites when traveling abroad. Such websites often restrict all access from abroad, for security or other reasons, which can be very inconvenient.

Based on the layer at which the VPN service provider's interchange VPN reachability information with customer sites, VPNs are classified into three types: layer 1 VPN (L1VPN), layer 2 VPN (L2VPN), and layer 3 VPN (L3VPN) [39, 40, 50, 51]. While L1VPN technology is under development, the other two technologies are mature and have been widely deployed. Also, based on their networking requirements, enterprises can connect their corporate locations together in many different ways. These networking services can typically be viewed from three perspectives, which are demarcation point (or enterprise/service provider handoff), the local loop (or access circuit), and the service core. Choosing layer 2 or layer 3 VPN will make a different impact on these three network services [52].

In an L2VPN, the service provider's network is virtualized as a layer 2 switch whereas it is virtualized as a layer 3 router in an L3VPN [39]. In the former, the customer sites are responsible for building their own routing infrastructure. Put differently, in an L3VPN, the service provider participates in the customer's layer 3 routing, while in an L2VPN it interconnects customer sites using layer 2 technology.

As listed in Tables 2.1 and 2.2, both Layer 2 and Layer 3 services have their advantages and disadvantages. These are basically related to the differences of router and switch in computer networking; some of them are highlighted in Table 2.3.

---

[5]You can also set up your own VPN to safely access your secure home network while you are on the road.

Table 2.1 Layer 2 VPNs: advantages and disadvantages

Advantages
Highly flexible, granular, and scalable bandwidth
Transparent interface—no router hardware investment is required
Low latency - switched as opposed to routed
Ease of deployment—no configuration required for new sites
Enterprises have complete control over their own routing
Disadvantages
Layer 2 networks are susceptible to broadcast storms—due to no router hardware
No visibility from the service provider— monitoring services can be difficult
Extra administrative overhead of IP allocations—because of flat subnet

Table 2.2 Layer 3 VPNs: advantages and disadvantages

Advantages
Extremely scalable for fast deployment
Readiness for voice and data convergence
"any to any" connectivity—a shorter hop count between two local sites
Enterprises leverage the service provider's technical expertize for routing
Disadvantages
Increased costs—due to requiring customer router hardware
Class of service and quality of service usually incur additional fees
IP addressing modifications would have to be submitted to the service provider

2.3.3 Virtual Sharing Networks

VPNs and overlays are not the only types of virtual networks implemented so far; there exist other networks that do not fall into these two categories. Virtual *local area networks* (Virtual LAN's) are examples of these networks. While properly segmenting multiple network instances, such technologies commonly support sharing of physical resources among them. The term *virtual sharing networks* (VSNs) has recently been suggested for these types of networks [39].

Originally defined as a network of computers located within the same area, today LANs are identified by a single *broadcast domain* in which the information broadcasted by a user is received by every other user on that LAN while it is prevented from leaving the LAN by using a router. The formation of broadcast domains in LANs depends on the physical connection of the devices in the network. Virtual LANs (VLANs) were developed to allow a network manager to logically segment a LAN into different broadcast domains. Thus, VLANs share the same physical LAN infrastructure but they belong to different broadcast domains. Since it is a logical, rather than a physical, segmentation, it does not require the workstations to be

Table 2.3 Router versus switch

	Router	Switch
Definition	A router is a network device that connects two or more networks together and forwards packets from one network to another	A switch is a device that connects many devices together on a computer network. It is more advanced than a hub
OSI Layer	Network Layer (L3) devices	Data Link Layer. Network switches are L3 devices
Data form	Packet	Frame (L2 switch)/Frame and Packet (L3 switch)
Address used for data transmission	IP address	MAC address
Table	Stores IP addresses in routing table and keeps them on its own	A network switch stores MAC addresses in a lookup table
Transmission type	At initial level broadcast then unicast and multicast	First broadcast; then unicast and multicast as needed
Routing decision	Takes faster routing decision	Takes more time for complicated routing decision
Used to connect	Two or more networks	Two or more nodes in the same or different network

physically located together. They can be on different floors of a building, or even in different buildings. Further, broadcast domain in a VLAN can be defined without using routers; instead, bridging software is used to define which workstations belong to the broadcast domain, and routers are only used to communicate between two VLAN's.

The sharing and segmentation concept of the VLAN can be generalized to a broader set of networks, collectively called *virtual sharing networks* (VSNs). The key requirement for such networks is to share a physical infrastructure while being properly segmented [39]. For example, a large corporate may have different networks with specific permission for guests, employees, and administrators, yet all sharing the same access points, switches, router, and servers.

2.3.4 Relation Between Virtual Networks

A virtual network can be considered "virtual" from different perspectives, so its type may change simply by changing the perspective. An overlay network is virtual as it is separated from the underlying physical network; a VPN is virtual since it is distinct from the public network; VSNs are virtual because multiple segmented networks share a same physical infrastructure. With these views, VPN can be considered an overlay network, as the tunnels used for connection are separate and external to the private network and used to extend the functionality and accessibility of the primary physical network. Likewise, overlay networks sharing the same underlay become VSN. However, it should be noted that overlay, VPN, and VSN respectively emphasis on new services, connectivity, and resource sharing.

In summary, network virtualization is an overlay; that is, to connect two domains in a network, it creates a tunnel through an existing network rather than physically connecting them. It saves administrators from having to physically wire up each new domain connection; especially, they need not change what they have already done; they make changes on top of an existing infrastructure.

While NV creates tunnels through a network, the next step to automate the network is to put services, such as firewall, on tunnels. This is what NFV offers, and is explained in the following section.

2.4 Network Functions Virtualization

There are increasing variety of proprietary hardware appliances to launch different services in telecommunication networks. Launching a new network service yet often requires another appliance, implying further space and power to accommodate these boxes, in addition to increased integration and deployment complexity. Further, as innovation accelerates, lifecycles of these hardware-based appliances becomes shorter and shorter, meaning that the return on investment is reduced. The above problems could be addressed if the services are run in *software*. Thus, enlightened by virtualization, the following question would arise: If administrators can set up a virtual machine by a click, why shouldn't they launch a service in a similar fashion?

Network functions virtualization[6] (NFV) decouples *network functions* from proprietary hardware appliances, to overcome the above deficiencies (Fig. 2.4). It offers a new way to architect and implement layer 4 through layer 7 network functions in order to run computationally intensive network services in software, that can be moved to standard hardware. Through decoupling layer 4–7 network functions, such as firewall, intrusion detection, and even load balancing, from proprietary hardware appliances, and implementing them in software, NFV offers a cost effective, and more efficient way to design, deploy, and manage networking services. Network functions virtualization is targeted mainly at the carrier or service provider market, and it enables operators to [53, 54]:

---

[6]The Network Functions Virtualization Industry Specification Group (NFV ISG) was initiated under the auspices of the European Telecommunications Standards Institute (ETSI). NFV ISG first met in January 2013, and will sunset two years after [53]; it included over 150 companies in 2013. The NFV ISG objective is not to produce standards but to achieve industry consensus on business and technical requirements for NFV. A more detailed version of [53] is expected to be released in the second half of 2014.

- **Reduce CapEx**: NFV reduces the need to purchase purpose-built hardware by using commercial off-the-shelf hardware which is typically less expensive than purpose-built, manufacturer-designed hardware. By shifting more components to a common physical infrastructure, operators save more. Also, through supporting pay-as-you-grow models, which eliminate wasteful overprovisioning, operators can save even more.
- **Reduce OpEx**: NFV reduces space, power, and cooling requirements of equipment since all services utilize a common hardware. Further, it simplifies the roll out and management of network services as there is no need to support multiple hardware models from different vendors.
- **Accelerate time-to-market**: NFV reduces the time required to deploy new network services. This in turn improves return on investment of new services. In addition, by reducing time-to-market, it lowers the risks associated with rolling out new services to meet the needs of customers and seize new market opportunities.
- **Increase flexibility**: NFV simplifies the addition of new applications and services as well as the removal of existing ones, to address the constantly changing demands and evolving business models. It supports innovation by enabling services to be delivered in software that can run on a range of industry-standard server hardware.

It should be noted that operators need to evolve their infrastructures as well as their operations/business management practices to fully benefit from NVF. The virtualized telecommunications network infrastructure requires more stringent reliability, availability, and latency in comparison to the cloud as it is currently used in the IT world [4].

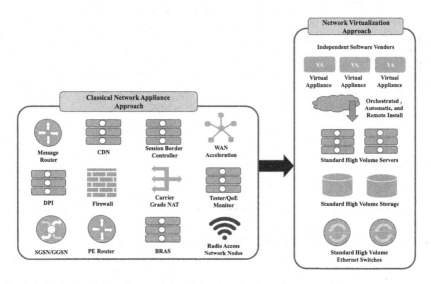

Fig. 2.4 The ETSI vision of network functions virtualization [3]

2.4.1 What to Virtualize?

In the mobile network, i.e., evolved packet core (EPC), IP multimedia subsystem (IMS), and RAN, we can, for example, virtualize mobility management entity (MME), serving gateway (SGW), packet data networks gateway (PGW), radio network controller (RNC), and base stations network functions. A virtualized EPC (vEPC) automates the authentication and management of subscribers and their services, whereas a virtualized IMS (vIMS) can deliver a portfolio of multimedia services over IP networks. The EPC and IMS network functions can be unified on the same hardware pool. Base station functions, e.g., PHY/MAC/network stacks that handle different wireless standards (2G, 3G, LTE, etc.) can share the centralized hardware resources and achieve dynamic resource allocation [53].

In general, the benefits of virtualizing network functions fall into two main categories, i.e., *cost saving* and *automation* gains and these benefits vary from system to another [4]. An assessment of the NFV benefits is shown in Fig. 2.5. While in many cases operators will benefit from virtualizing network functions, there are a few exceptions. For example, virtualizing high-performance routers or Ethernet switches is not expected to result in cost saving. Further, virtualizing products primarily focused on packet forwarding may or may not be cost effective, depending on their deployment and the ratio of control versus data plane traffics [4]. Nevertheless, even when there are no cost savings, virtualization could be justified, at times, by automation gain.

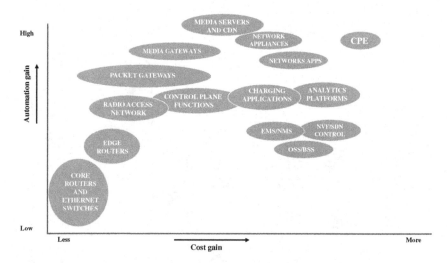

Fig. 2.5 Network functions virtualization: an assessment of the benefits [4]

2.5 Wireless Virtualization

As a natural extension of wired network virtualization, wireless networks virtualization is motivated by the observed benefits of that in wired networks. However, while virtualization of wired networks and computing systems has become a trend, much less virtualization has occurred in infrastructure based wireless networks [55]. Yet, the idea of virtualizing wireless access has recently attracted substantial attention in both academia and industry. It is one of the frontier research areas in computer science [39, 55, 56].

Wireless virtualization may refer to wireless access virtualization, wireless infrastructure virtualization, wireless network virtualization, or even mobile network virtualization [39, 56]. It is about the abstraction and sharing of wireless resources and wireless network devices among multiple users while keeping them isolated. Wireless resources may include low-level PHY resources (e.g., frequency, time, and space) or wireless equipment (e.g., a base station (BS)[7]), a network device (e.g., a router), a network, or a client hardware (e.g., wireless NIC). Thus, similar to wired network virtualization, wireless network virtualization software may reproduce logical channel and logical RAN (L1) in addition to logical switches and logical routers (L2–L3).

The motivations for virtualizing wireless networks are very similar, but not limited, to those of wired networks. First, as an extension of wired network virtualization, wireless virtualization can potentially enable separation of traffic to increases flexibility (e.g., in terms of QoS), improve security, and facilitate manageability of networks. Powerful network management mechanisms are particularly important in emerging heterogeneous networks. Second, it has a great potential to increase the utilization of wireless networks. This is important from both infrastructure and spectrum virtualization points of view. The former opens up the doors for the concept of infrastructure as a service (IaaS) so that one operator can use its own or other operators underutilized equipment (e.g., BSs on the outskirts) in the congested sites, for example in downtown. Spectrum virtualization can also provide better utilization; it may even bring more gain and is more valuable as spectrum is a scarce resource. Third, by decoupling the logical and physical infrastructures, wireless virtualization promotes mobile virtual network operators (MVNOs[8]). This allows decoupling operators from the cost of infrastructure ownership (capital and operation expenditures). Fourth, wireless virtualization provides easier migration to newer products and will likely support the emergence of new services. Last but not the least, it is a key enabler for cloud radio access network, which is expected to help operators reduce TCO and become greener.

---

[7] A single physical BS can be abstracted to support multiple mobile operators and allow individual control of each RAN by having a separate vBS configured for each operator.

[8] MVNOs [57] are a new breed of wireless network operators who may not own the wireless infrastructure or spectrum, but give a virtual appearance of owning a wireless network. Basically, MVNOs resell the services of big operators, usually lower prices and with more flexible plans. Virgin Mobile is an example for MVNO.

Depending on the type of the resources being virtualized and the objective of virtualization, three different generic frameworks can be identified for wireless virtualization [56]:

1. **Flow-based virtualization** deals with the isolation, scheduling, management and service differentiation between *traffic flows*, streams of data sharing a common signature. It is inspired by the flow-based SDN and network virtualization but in the realm of wireless networks and technologies. Thus, it requires wireless-specific functionalities such as the radio resource blocks scheduler to support quality of service (QoS) and service-level agreement (SLA) over the traffic flows.

2. **Protocol-based virtualization** allows to isolate, customize, and manage multiple wireless protocol stacks on a single radio hardware, which is not possible in flow-based virtualization. This means that MAC and PHY resources are being virtualized. Consequently, each tenant can have their own MAC and PHY configuration parameters while such a differentiation is not possible in a flow-based virtualization. The wireless network interface card (NIC)[9] virtualization [60, 61] where IEEE 802.11 is virtualized by means of the 802.11 wireless NIC, falls into this category.

3. **RF front end and spectrum-based virtualization** is the deepest level of virtualization which focuses on the abstraction and dynamic allocation of the spectrum. Also, it decouples the RF frontend from the protocols and allows a single front end to be used by multiple virtual nodes or a single user to use multiple virtual frontends. The spectrum allocation in the spectrum-based virtualization differs from that of the flow-based virtualization for its broader scope and potential to use noncontiguous bands as well as the spectrum allocated to different standards.

As noted, the depth of virtualization is different in these three frameworks and they are complementary to each other. From an implementation perspective, the flow-based virtualization is the most feasible approach with immediate benefits. It connects virtual resources and provides a more flexible and efficient traffic management. In all three cases, a flow-based virtualization is required to integrate the data. However, in the flow-based approach, the depth of virtualization is not sufficient for more advanced wireless communication techniques, such as the *coordinated multipoint* transmission and reception [56, 62, 63].

As a potential enabler for future radio access network, wireless virtualization is gaining increasing attention. However, virtualization of wireless networks, especially efficient spectrum virtualization is far more complicated than that of a wired network. It faces some unique challenges that are not seen in wired networks and data centers. Virtualization of the wireless link is the biggest challenge in this domain [56, 64, 65]. Some other key issues in wireless virtualization are:

---

[9]By means of a wireless NIC, which is basically a Wi-Fi card, a computer workstation can be configured to act as an 802.11 access point. As a result, 802.11 virtualization techniques can be applied to the 802.11 wireless NIC. Virtualization of WLAN, known as VirtualWi-Fi (previously MultiNet [58, 59]) is a relatively old technology. It abstracts a single WLAN card as multiple virtual WLAN cards, each to connect to a different wireless network. Therefore, it allows a user to simultaneously connect his machine to multiple wireless networks using a single WLAN card.

- **Isolation**: *Isolation* is necessary to guarantee that each operator can make independent decision on their resources [66]. Also, since resources are shared in a virtualized environment, there must be effective techniques for ensuring that the resource usage of one user has little impact on others. In wired networks, this may only occur when every user is not provided with distinct resources, mainly due to resources insufficiency. *Overprovisioning* can solve the issue in such cases. It is not, however, a viable solution in wireless virtualization because *spectrum*, the key wireless resource, is scarce. To fulfill such requirements, sophisticated dynamic resource partitioning and sharing models are required.
- **Network management**. Wireless networks are composed of various radio access technologies (RATs), e.g., 3G, 4G, and Wi-Fi. Similarly, a single wireless device is capable of accessing to multi-RAT. In such a multi-RAT environment, resource sharing is not straightforward. In contrast to network virtualization technologies which are mainly based on Ethernet, wireless virtualization must penetrate deeper into the MAC and PHY layers. Further, even in a single RAT environment, slicing and sharing is not easy because wireless channels are very dynamic in nature and an efficient slicing may require dynamic or cognitive spectrum sharing methods [67]. Hence, *dynamic network virtualization* algorithms must be considered.
- **Interference**: Wireless networks are highly prone to interference and their performance is limited by that. Interference is out there, particularly, in dense, urban area. This must be considered in slicing radio resources since it is not easy to isolate and disjoint subspaces. Especially, in a multi-RAT environment, if different spectrum bands of various RATs are shared and abstracted together, interference becomes even a bigger issue because interference between different RAT needs to be taken into account too. For example, a slice from WiFi unlicensed spectrum could be assigned to an LTE user, causing unforeseen interference between LTE and WiFi networks.
- **Latency**: [66] Current wireless standards impose very strict latency, in order to meet real-time applications requirement [68]. This mandate 5–15 ms round-trip latency in layer 1 and layer 2 of today's wireless standards and will be more stringent in the next generation (5G) [69].

There are also other concerns like synchronization, jitter [70], and security [71].

2.5.1 State of the Art in Wireless Virtualization

Technical advances will be discussed in Sect. 7.6. Here we consider the state of research in this field.

2.6 Cloud Computing

Cloud computing refers to delivering computing resource as a service over the Internet, on a pay-as-you-go pricing. This type of computing relies on sharing a pool of physical and/or virtual resources, rather than deploying local or personal hardware and software. The name "cloud" was inspired by the cloud symbol that has often used to represent the Internet in diagrams. Thanks to cloud computing, wherever you go your data goes with you.[10] Today, many large and small businesses use cloud computing, either directly or indirectly. The big players in the cloud space are: Amazon (AWS), Microsoft (Azure), Google (Google Cloud Platform), and Rackspace (OpenStack).

But, what explains the wide use of cloud computing among businesses? Costs reduction is probably the main driver. Cloud computing helps businesses reduce overall IT costs in multiple ways. First, cloud providers enjoy massive *economies of scale*. Effective use of physical resources due to *statistical multiplexing* brings prices lower, 5–7 times [72]. Then, multiple pricing models, especially, pay-per-use model, allow customers to optimize costs.[11] Cloud computing brings down IT labor costs and gives access to a full-featured platform at a fraction of the cost of traditional infrastructure. Universal access is another advantage of cloud computing. It allows remote employees to access applications and work via the Internet. Other important benefits include a choice of applications, flexible capacity, up to date software, potential for greener communication, and speed and agility. With flexible capacity, the organizations need not be concerned about over/under-provisioning for a service. When there is a load surge they can enjoy the infinite computing capacity on demand, and get results as quickly as their program scales, since there is no price difference in using 1000 servers for an hour or one server for 1000 h [72].

2.6.1 Cloud Services Models

Broadly speaking, public cloud services are divided into three categories: Infrastructure as a Service (IaaS), Platform as a Service (PaaS), and Software as a Service (SaaS). In general, X as a Service (XaaS) is a collective term used to refer to any

---

[10]Today, many people actually use cloud even before they knew it. The photos that you store on your social networking sites (e.g., Facebook) or any kind of file you store and view in online file storage sites (Dropbox, Google Drive, etc.) are stored on their servers, which can be accessed from anywhere by simply logging in with your account information. In addition, you may have used or heard about Google Docs, where you can create, store, and share documents (Word) and spreadsheets (Excel) on their server, once you have a Gmail id. It is also the same business model for emails services (Gmail, Yahoo mail, etc.) as you can log in to access your emails anywhere you want.

[11]While most major cloud service providers such as Azure, AWS, and Rackspace have an hourly usage pricing model, since March 2014 Google Compute Engine has started providing a per-minute pricing model.

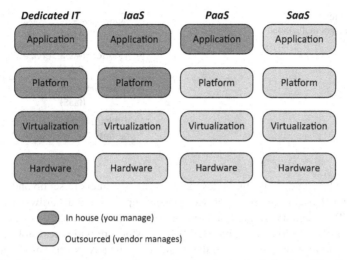

Fig. 2.6 Dedicated hosting versus cloud computing (purchasing IaaS, PaaS, and SaaS)

services that are delivered over the Internet, rather than locally. XaaS presents the essence of cloud computing and new variants of XaaS emerge regularly.[12] Yet, the three basic models (IaaS, PaaS, and SaaS) suffice for a proper understanding of cloud computing. This three service models form a service growth model, from IaaS through Paas to SaaS, as illustrated in Fig. 2.6, in which the following layers can be identified [73]:

- **Application** denotes the software for the customer.
- **Platform** includes runtime environment (e.g., .NET, PHP), middleware, and operating system in which software is run.
- **Virtualization** refers to the virtualization software (hypervisor) which creates multiple virtual environments based on the physical hardware.
- **Hardware** is the equipment (servers, storage, and network resources).

As can be seen in Fig. 2.6 and more clearly in Fig. 2.7, the first growth phase is the use of IaaS. IaaS, PaaS, and SaaS are different logical layers in the stack, as visualized in Fig. 2.7. The level of abstraction/control increases as we move up/down the stack. Here is a brief explanation of each service model.

- **Infrastructure as a Service (IaaS)**: In this case, computing resources (compute, storage, and network) are exposed as a capability and the clients put together their own infrastructure. For example, they decide on the operating system, the amount of storage, and the configuration of network components such as firewalls. The clients do not own, manage, or control the underlying infrastructure; instead, they rent it, as a service. As can be seen in Fig. 2.6, the hardware and virtualization

---

[12]Other examples of XaaS are Storage as a Service (SaaS), Desktop as a Service (DaaS), Network as a Service (NaaS), and Monitoring as a Service (MaaS).

Fig. 2.7 Different service models or layers in the cloud stack

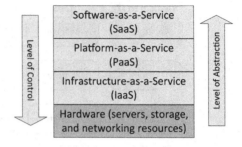

move to the cloud. This eliminates the need for customers to set up and maintain their own physical resources. Service provider supplies virtual hardware resources (e.g., CPU, memory, storage, load balancer, virtual LANs, etc.). An example is Amazon elastic cloud compute (EC2) [74], which provides resizable compute capacity along with the needs of the customers. The pay-per-use model makes it possible to not have to make more expenditure than is strictly necessary.

- **Platform as a Service (PaaS)**: In this solution, programming platforms and developing tools (such as Java, .NET, or Python) and/or building blocks and APIs for building and running applications in the cloud are provided as a capability. The customer has control over the applications and some of the configuration of the platform environment but not over the infrastructure; this is the main difference between PaaS and IaaS. Hence, unlike IaaS where users select their operating system, application software, server size, etc., and maintain complete responsibility for the maintenance of the system, with PaaS operating system updates, versions, and patches are controlled and implemented by the vendor.
Facebook is probably the most well-known PaaS. Web hosting is another example of PaaS, where web hosting provider provides an environment with a programming language such as PHP and database options in addition to hypertext transfer protocol (HTTP) which allow a personal website to be developed. Some of the biggest names in PaaS include Amazon Elastic Cloud Computing, Microsoft Azure, and Google App Engine, and Force.com [75, Chap. 8].

- **Software as a Service (SaaS)**: With SaaS, the customer uses applications, both general (such as word processing, email, and spreadsheet) and specialized (such as customer relationship management and enterprise resource management) that are running on cloud infrastructure. The applications are accessible to the customers, at any time, from any location, and with any device, through a simple interface such as a web browser. As it is shown in Fig. 2.6, all layers are outsourced in SaaS. This is the ultimate level of abstraction and the consumer only needs to focus on administering users to the system. Then again, the users can influence configuration only in a limited manner, e.g., the language setups and look-and-feel settings [73].

2.6.2 Types of Clouds

Originally synonymous with public clouds, today cloud computing breaks down into three primary forms: *public, private*, and *hybrid* clouds.[13] Each type has its own use cases and comes with its advantages and disadvantages.

Public cloud is the most recognizable form of cloud computing to many consumers. In a public cloud, resources are provided as a service in a virtualized environment, constructed using a pool of shared physical resources, and accessible over the Internet, typically on a pay-as-you-use model. These clouds are more suited to companies that need to test and develop application code and bring a service to market quickly, need incremental capacity, have less regulatory hurdles to overcome, are doing collaborative projects, or are looking to outsource part of their IT requirements. Despite their proliferation, a number of concerns have arisen about public clouds, including security, privacy, and interoperability. What is more, when internal computing resources are already available, exclusive use of public clouds means wasting prior investments. For these reasons, private and hybrid clouds have emerged, to make the environments secure and affordable.

Private clouds, in a sense, can be defined in contrast to public clouds. While a public cloud provides services to multiple clients, a private cloud, as the name suggests, ring-fence the pool of resources, creating a distinct cloud platform that can be accessed only by a single organization. Hence, in a private cloud, services and infrastructure are maintained on a private network. Private clouds offer the highest level of security and control. On the other hand, they require the organization to purchase and maintain its own infrastructure and software, which reduces the cost efficiency. Besides, they require a high level of engagement from both management and IT departments to virtualize the business environment. Such a cloud is suited to businesses that have highly critical applications, must comply with strict regulations, or must conform to strict security and data privacy issues.

A hybrid cloud comprises both private and public cloud services. Hence, it is suited to companies that want the ability to move between them to get the best of both the worlds. For example, an organization may run applications primarily on a private cloud but rely on a public cloud to accommodate spikes in usage. Likewise, an organization can maximize efficiency by employing public cloud services for nonsensitive operations while relying on a private cloud only when it is necessary. Meanwhile, they need to ensure that all platforms are seamlessly integrated. Hybrid clouds are particularly well suited for E-commerce since their sites must respond to fluctuating traffic on a daily and seasonal basis. On the downside, the organization has to keep track of multiple different security platforms and ensure that they can communicate with each other. Regardless of its drawbacks, the hybrid cloud appears to be the best option for many organizations.

---

[13]Some add a fourth type of cloud, called *community cloud* [75]. It refers to an infrastructure that is shared by multiple organizations and supports a specific community. The healthcare industry is an example of an industry that is employing the community cloud concept.

Table 2.4 Cloud computing: benefits and risks

Cloud type	Benefits	Drawbacks
Public	• Low investment in the short run (pay-as-you-use)	• Security: multi-tenancy and transfers over the Internet [76]
	• Highly scalable	• Privacy and reliability [76]
	• Quicker service to market	
Private	• More control and reliability	• Higher cost: heavy investment in hardware, administration and maintenance
	• Higher security	• Must comply with strict regulations
	• Higher performance	
Hybrid	• Operational flexibility: can leverage both public and private cloud	• Security, privacy, and integrity concerns
	• Scalability: run bursty workloads on the public cloud	
	• Cost effective	

In Table 2.4, we enlist the main benefits and risks associated with each type of clouds. Understandably, security is one of the main issues in cloud computing. There are many obstacles as well as opportunities for cloud computing. Availability and security are among the main concerns [72, 77].

2.6.3   Virtualization Versus Cloud Computing

By now the reader should have realized the connection between virtualization and cloud computing. Broadly speaking, these two technologies share a common bond: they are both meant to increase efficiencies and reduce costs. They are quite different though. Virtualization is one of the elements that forms cloud computing. It is the software that manipulates hardware, while cloud computing is a service that results from that manipulation [78].

Observing that cloud computing is built on a virtualized infrastructure, one can deduct that if an organization have already invested in virtualization, they may bring in cloud to further increase the computing efficiency. Then, the cloud could work on top of the current virtualized infrastructure; it also helps in the delivery of current network as a service. Put differently, cloud computing makes use of virtualized resources at a different level, where the resources can be accessed as a service, and in an on-demand manner. Conversely, any organization considering adoption of a private cloud must work on virtualization, too.

Organizations can improve the computing resources efficiency through virtualization; however, they cannot get rid of provisioning. An administrator is still required to provision the virtual machines for the users. Cloud computing removes the need for manual provisioning. It offers a new way for IT services delivery by providing a customer interface to automated, self-service catalogs of standard services, and by using autoscaling to respond to increase or decrease in users demand [79].

2.7 Summary

Network overlay (using encapsulation and tunneling techniques) is one way to implement virtual networks. This approach is network agnostic, but it cannot reserve resources such as bandwidth. In addition, it does not guarantee service quality, hence it can result in degraded application performance.

Using software-defined networking (SDN) is another way to implement network virtualization. For example, one can define the virtual networks in the flow tables of the SDN switches. The SDN approach to network virtualization overcomes the limitations of the above approach. It also brings the ability to do more granular traffic routing and to gather more intelligence about the infrastructure.

Chapter 3
Software-Defined Networks Principles and Use Case Scenarios

3.1 Introduction

Over the next decade, network traffic is expected to grow exponentially, demanded by applications such as video and machine to machine applications and made possible by broadband technologies such as long-term evolution (LTE). This has made new challenges for network operators and service provider: to design and architect networks for reducing TCO, and improving average revenue per user (ARPU) and customer retention.

The software defined networking (SDN) paradigm provides a promising way to address these challenges with four fundamental principles: (1) decoupling of control plane and forwarding plane, which scaling and reduces TCO, (2) open APIs, which make the network programmable and hence flexible. With open APIs, the rigid architecture of current networks will be replaced with flexible and evolving architectures for service velocity and innovation, (3) network virtualization, which provides resource optimization, and (4) control plane signaling mechanism/protocol, which intelligently manages the forwarding plane.

Among different use cases of SDN, access/aggregation domains of public telecommunication networks and data center domains are two important use cases. Access/aggregation network domains include mobile backhaul, where the virtualization principle of SDN could play a critical role in isolating different service groups and establishing a shared infrastructure that could be used by multiple operators. In Sect. 3.5 we will focus on some of the key features of SDN from which a future access/aggregation domain could benefit.

Applications of SDN in data centers could be manifold. Two of the most interesting aspects are load balancing and dynamic resource allocation (hence increased network and server utilization) and the support of session/server migration through dynamic coordination and switch configuration. The decoupling principle of SDN could provide the possibility to develop data center solutions based on commodity switches instead of typically high-end expensive commercial equipments.

© Springer International Publishing AG 2017
M. Vaezi and Y. Zhang, *Cloud Mobile Networks*, Wireless Networks,
DOI 10.1007/978-3-319-54496-0_3

In the following section, we discuss SDN technology and business drivers, architecture and principles, and use case scenarios. We provide insights on design implementation consideration of SDN in carrier-grade networks, which are characterized by some specific challenges including architectural requirements for scalability and reliability. This paper is the first work to investigate practical applications and implementation concerns of SDN from the production network's perspective.

3.2 SDN Drivers

SDN is gaining significant interests from both academia and industry. In the following, we motivate the needs for this new network architecture from both technology and business drivers' perspectives.

3.2.1 Technology Drivers

Scaling Control and Data Plane

Over the next decade, it is expected that exponential growth of traffic would continue as we see broader adoption of new broadband technologies such as FTTH and advanced LTE. This bandwidth growth is fueled by applications such as video and machine to machine (M2M) application, with varying traffic characteristics: Video applications being more bandwidth consuming with a need for scaling data plane and M2M applications being more of signaling consumed application with a need for scaling control plane. These different types of traffic in the next generation networks would pose independent scaling of control and data plane to meet next generation traffic growth and demands.

Service Velocity and Innovation

As we see application world evolving rapidly with over-the-top, user generated content, etc., it is becoming imperative that network be programmable for service providers to use network as a platform to expand their service models to include third-party applications, truly move from self-contained to ecosystem-based business models. This also means, rapid introduction of not only own services so-called walled garden application, but also including third-party applications from other content providers.

Network Virtualization

Currently, service providers deploy an overlay network for each type of application. For example, they have one network for wireline application, another one for wireless and another for business applications. As we move toward converged networks, the goal is to design one network for many services by slicing this physical network into multiple virtual networks, one for each type of application.

3.2.2 Business Drivers

Service providers are driven by three-dimensional problems while planning to transform from simple connectivity providers to experience providers. These three-dimensional business drivers are: TCO reduction, increased ARPU, and improved customer retention. These business drivers are primarily connected and influence the way networks are architected and designed. While mapping these business drivers to technology drivers, scaling and virtualization are primarily required to address reducing TCO, while service velocity and innovation is required to address increasing ARPU and customer retention.

In following section, we discuss SDN architecture and principles, show how SDN addresses the service providers technology drivers to help focus on addressing their business drivers.

3.3 Architecture and Principles

Given the significant interest because of the key drivers, an increasing amount of attention has been raised on SDN. However, until now, there has been no consensus on the principles and concepts of SDN. In the following, we take the initiative to summarize the key aspects of SDN from service providers' perspectives. The four main principles of software-defined networking in service provider networks are: *split architecture, open APIs, network virtualization*, and *control plane signaling protocol*. Figure 3.1 illustrates the four principles.

Fig. 3.1 Four principles of SDN

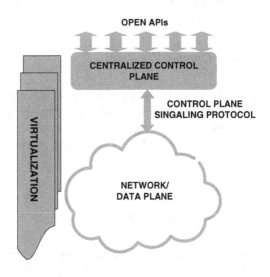

3.3.1 Split Architecture

The separation and centralization of the control plane software from the packet-forwarding data plane is one of the core principles of SDN. This differs from the traditional distributed system architecture in which the control plane software is distributed across all the data plane devices in the network.

The advantages of centralized control plane are as follows:

- **New revenue opportunities**: Centralized network information enables deployment of new services and applications that are not possible or are very difficult in traditional distributed networks
- **High service velocity**: Software-based view of the network enables deployment of new services much faster compared to upgrading a whole networking device as it is done by current vendors.
- **Lower Capex**: The separation of control and data planes enable independent and parallel optimizations of the two planes. We envision that this path will lead to highly specialized and cost-effective high-performance packet forwarding data plane devices and control plane servers, thus reducing the capex of service providers significantly.
- **Lower Opex**: Operations and maintenance (OAM) is one of most important aspects for network administrators and operators. Maintenance operations such as in-service software upgrades (ISSU) of network devices are an expensive and time-consuming proposition. SDN plays a very important role is optimizing the cost and effort for network OAM. Since network information, applications, and services are concentrated at a centralized location in SDN, operations such as ISSU become much easier and cost-effective by avoiding individual software upgrades at multiple device locations prevalent in current distributed networks.

3.3.2 Open-APIs

Open APIs enable developers to exploit the centrally available network information and the mechanisms to securely program the underlying network resources [80]. They not only enable rapid development and deployment of both in-house and third-party applications and services but also provide a standard mechanism to interface with the network.

3.3.3 Virtualization

Network virtualization is an important area of research in service provider networks [81–83]. Network virtualization, analogous to server virtualization, enables precious network resources to be used by multiple network operators while maintaining network availability, reliability, and performance. Network virtualization paves the path for converged networks and higher return on investment of network resources. In order to achieve a complete virtual network, virtualization of resources is required at the control plane, control channel, and the data plane levels.

3.3.4 Control Plane Signaling Protocol

A secure and extensible control plane signaling protocol is an important component for the success of SDN. It should enable efficient and flexible control of network resources by the centralized control plane. OpenFlow [84] is a well-known control plane signaling protocol that is standardized and increasingly being made extensible and flexible.

3.4 Use Case Scenarios

The four principles of SDN can have concrete representations in different network contexts and applications. In this section we discuss the application of SDN for different usecase scenarios. We specifically describe access/aggregation in mobile backhaul networks, and inter-data center WAN scenarios and discuss how SDN could be applied in these usecases.

3.4.1 SDN in Mobile Backhaul Networks

Figure 3.2 shows the state-of-the-art access/aggregation network architecture used by most carriers today. These networks primarily use Ethernet to provide triple-play residential services as well as business services as defined by the Metro Ethernet Forum. However, the trends driving the future of access/aggregation networks are determined by the desire of carriers to have a unified and simplified control and management plane to operate multi-service and multi-tenant networks.

The current access/aggregation networks are divided in multiple dimensions—different administrative domains, different transport technologies, and different control and management planes, among others. Usually, a carrier has separate departments and competencies to manage fixed networks and mobile backhaul networks. This is a result of legacy as well as the different technologies and control planes used in both these networks. Similarly on the transport side, optical networks (SDH/SONET) and IP/Ethernet networks have quite different control and management planes. We envision that extending MPLS from the core to the backhaul can provide a unifying and simplifying effect on carrier networks. A unified MPLS-based solution

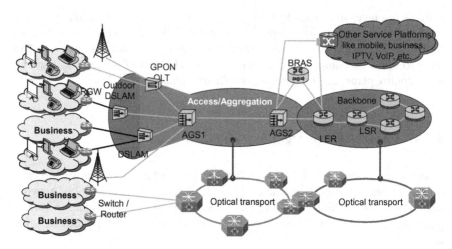

Fig. 3.2 Mobile backhaul networks (access/aggregation)

simplifies network control and management by abstracting various underlying technologies such as L1/TDM, L2/ATM/Ethernet/HDLC, and L3/IP. Further, it can
provide the inherent advantages of MPLS such as scalability, security, and traffic
engineering to the access/aggregation networks. Ideas for a combined MPLS-based
solution for both the core and the aggregation networks are gaining wider attention
lately (see seamless MPLS [85]). Introduction of MPLS in the aggregation network
enables independent evolution and optimization of transport and service architecture, which matches with the SDN philosophy and architecture very closely. The
need for a unified and simplified control and management plane is further increased
by the requirements to support multiple services and multiple operators on the same
physical network.

The introduction of SDN into the mobile backhaul networks not only unifies and
simplifies network architecture but also enables multi-service deployment through
Open-APIs and multi-tenant deployment through advanced mechanisms such as
network virtualization. Open-APIs enable in-house as well as third party development
and deployment of services by enabling the services to program the underlying
networks as needed while enforcing security and network policies defined by the
operator. Network virtualization enables "slicing" of dataplane, control channel,
and centralized control resources to be used by multiple operators using the same
underlying physical network.

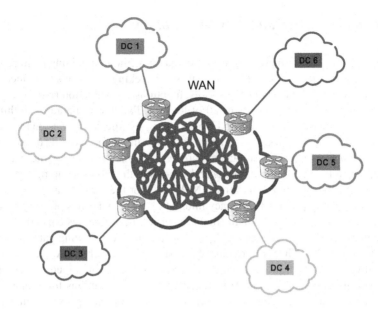

Fig. 3.3 Inter-data enter WAN

3.4.2 SDN in the Cloud

Data centers are an important area of interest for service providers. Applications such as "cloud bursting" are bringing together private and public data centers like never before. Distributed data centers are becoming the rule rather than the exception for many enterprizes and service providers. In such scenarios, the inter-data center wide area network (WAN) providers become a very important component in efficient and cost-effective operation of distributed data centers. Figure 3.3 illustrates this. The static provisioning mechanisms used currently by operators to provision resources on the WAN will not only degrade the performance of distributed data centers but also increasing the cost of operation. Thus it is necessary that the inter-data center WAN evolves into a networking resource that can allocated and provisioned with the same agility and flexibility that is possible within data centers. SDN is a very good candidate to enable such agile and flexible WAN operations the data center operators are expecting. Centralized control of the WAN using a SDN controller, with its Open-APIs and virtualization mechanisms, not only provides a single unified control of the WAN resources but also enables on-demand provisioning of network resources for multiple clients based on their policies and SLAs, in a seamless fashion.

Another major advantage of a centralized SDN controller in the inter-data center WAN is that it facilitates unified control along with data center resources. Data center operators are increasingly demanding a single unified mechanism to manage all the data centers in their domain instead of separate, distributed management. SDN on the WAN is one of the mechanisms to achieve this.

3.4.3   SDN in NFV and Service Chaining

To meet the constantly increasing traffic demands while maintaining or improving average revenue per user (ARPU), operators are seeking new ways to reduce their costs. To this end, the concept of network functions virtualization (NFV) was formally introduced in 2012 by about 20 of the world's largest operators, within the ETSI consortium [86]. Since then, NFV is bringing significant impacts on the way to view, and plan for, network infrastructure evolution in the forthcoming decade. Currently, the NFV initiative has attracted more than 200 members and participants that are actively defining the requirements and validating the NFV technologies through proof of concepts and other initiatives. NFV calls for the virtualization of network functions currently provided by legacy middleboxes and gateways offering network services such as firewalls, content filters, intrusion detection systems (IDS), deep packet inspection (DPI), network address translation (NAT), serving/gateway GPRS support node (SGSN/GGSN), broadband remote access server (BRAS), session border controllers (SBCs), provider edge (PE) routers, etc. Using cloud technologies, NFV will allow consolidation of these legacy network functions traditionally run on dedicated hardware. This consolidation promises operating expenditure gains mainly achieved through improved management of network functions. Moreover, NFV promises capital expenditure gains by running on generic server hardware and by leveraging cloud technologies for resource sharing. At last, NFV will provide greater agility to introduce new network functions, resulting in faster time to market of new services. Service chaining is the well-known process of forwarding traffic through a desired set of network functions (i.e., services or middleboxes). For cost and efficiency reasons, operators try to avoid sending all the traffic through every possible service. Depending on the traffic type, service-level agreement (SLA), and other factors, a provisioning policy dictates an ordered set of services for each traffic flow to traverse. Also, for load balancing purposes, traffic is sent of one of the often many instances of the same service. Operators have often struggled with this problem as most of these services (legacy middleboxes) had different forwarding behavior ranging from being bump-in-the-wire services (i.e., not IP reachable) to re-writing packet headers for internal functionalities (making the use of tagging technologies impossible). To this end, the literature contains a great collection of proposals to solve service chaining for these legacy middleboxes and novel ideas based on SDN techniques [87, 88]. While NFV may dictate better design requirements in support of service chaining, the problem of efficiently sending the traffic through the desired set of vNFs remains an open issue. With NFV, network-function placement can be highly flexible, for example, network functions can be instantiated at servers with low load or reachable with less-congested links. This imposes new challenges on DC networks to dynamically steer incoming traffic to servers that host the target network functions. SDN can be a good solution for providing service chaining functionalities for NFVs. Using NFV, we can dynamically instantiate new instances of vNFs, and use SDN to steer traffic dynamically through it.

SDN, NV and NFV all are designed to address mobility and agility. NV and NFV add virtual tunnels and functions to the physical network and can work on existing networks while SDN changes the physical network and requires a new network construct with separated data and control planes.

3.4.4 SDN Based Mobile Networks

In cellular networks, SDN has been proposed to simplify the design and management of cellular data networks, and to enable new services. It addresses the scalability introduced by the centralized the data-plane functions in the packet gateway, including monitoring, access control, and quality-of-service functionalities. The gateways usually locate in the core network of SDN, forcing all the traffic to traverse to the core network even for traffic within the cellular network. This results in additional bandwidth consumption. The benefit of SDN in cellular network can be summarized to the following aspects.

- Flexible management of subscribers' policies. The SDN controller can maintain a database of the subscribers' information and construct the detailed packet forwarding rules according to the policies based on the subscribers' information. Note that the policy can be defined by multiple factors, e.g., subscriber ID, device type, access type.
- Diverse operations on packets at the data plane. The SDN switches can process packets based on any fields in the packet header. Thus, complex operations on packets can be delegated from specialized middleboxes to the set of distributed switches.
- Integrated control of radio resources, backhaul and core network resources. As the virtualization becomes more and more popular, it has been evident that the radio resources may have open interfaces controlled by the same SDN controller. Thus, cross-domain optimization is possible by combining the knowledge from multiple networks.

Below we discuss a few existing work on SDN based cellular network and provide an overview of the future directions in this space.

Reference [89] is a position paper that describes the initial use cases of SDN in cellular networks. It discussed a few opportunities in applying SDN in cellular networks. First, SDN provides fine-grained packet classifier and flexible routing, which can easily direct a chosen subset of traffic through a set of middleboxes. As a result, middleboxes will handle much less traffic, making them much cheaper. Second, Cellular network over requires fine-grained and fast packet counting to handle user mobility, changing channel conditions, and load balancing. The SDN switches can be installed with measurement rules dynamically. By adjusting these rules over time, the cellular provider can efficiently monitor traffic at different levels of granularity to drive real-time control loops on the SDN controller. Third, subscriber mobility can be handled by the SDN's direct control over routing. Routing across multiple base stains and different cellular technologies can be achieved by using the unified Openflow protocol. Finally, SDN enables different carriers to share the infrastructure to offer a complete virtual LTE network to their customers.

SoftCell [90] is an architecture that supports fine-grained policies for mobile devices in cellular core networks based on commodity SDN switches. It provides flexible ways to steer traffic through sequences of middleboxes based on subscriber attributes and application types. It aggregates traffic along multiple dimensions in order to reduce the size of the forwarding table. It support multiple levels of polices, including the application type, subscriber type, and the base station ID. The packet is classified at the base stations as soon as they enter the mobile network by the SDN switches. To handle mobility it keeps track of the movement of the users and updates the sequence of middleboxes when the user equipment moves within the network.

Different from SDN's normal use cases in the fixed network, SoftRAN [65] is a software defined radio access layer. It has a software defined centralized control plane for radio access networks that abstracts all base stations in a local geographical area as a virtual base station. The virtualized base station consists a hierarchy of multiple physical base stations, controlled by a central controller. SoftRAN has been implemented in LTE-sim. For future work, research is needed to evaluate the performance and scalability of SoftRAN in both the hardware and software approaches.

3.5 SDN Design Implementation Considerations

Deploying SDN in different network contexts may have distinct requirements. However, there are a few fundamental design issues to be considered across all different use cases. In this section, we present a discussion of three practical implementation requirements. For each topic, we discuss potential solutions on how to address them.

Unlike the traditional network architecture, which integrates both forwarding (data) and control planes on the same box, split architecture decouples these two and runs the control plane on servers which might be in different physical locations from the forwarding elements (switches) [84]. Such decoupling imposes new challenges to the implementation and deployment in a large scale. In particular, we focus on there challenges: scalability, reliability, and controller-to-switch connectivity below.

3.5.1 Scalability and High Availability

One of the main requirements for a centralized SDN controller to be deployed in a production network is its scalability and high availability. We propose that the key requirements for SDN scalability is that the centralized SDN controller can maintain high performance and availability with increasing network sizes, network events, and unexpected network failures. We focus our discussions on the controller scalability below.

To improve the controller scalability, a common practice is to deploy multiple controllers either for load balancing purposes or for back-up purposes. According to different purposes, we can employ different models to provide a more scalable design. In the following, we discuss three different models for scalability and high availability for centralized SDN controllers: Hot-Standby Model, Distributed Network Information Model, and a Hybrid Model. The last model incorporates important aspects of the previous two models. We argue the third model is more appropriate for complex network scenarios with carrier-grade scalability requirements.

- The *hot-standby model* is similar to the 1+1 high availability models of current off-the-shelf router and switches. In this model, a master controller is protected by a hot-standby model. The standby instance will take over the network control upon failure of the master controller. The advantage of this model is its simplicity. However, it may encounter performance bottleneck when the number of switches and the communication messages grow significantly.
- In contrast, the *distributed network information model* incorporates concepts employed in today's massive data centers for data scalability, availability, and consistency. In this model the network controller is a cluster of controllers. Each of the controllers control a different part of the network. The network information is replicated across multiple controllers for high availability and scalability. This model is designed for large network with hundreds or even thousands of switches. The disadvantage is the communication overheads between controllers. With a careful design communication protocol, the drawbacks could be overcome.
- Finally, the *hybrid model* is a combination of the previous two models where the network information is replicated for high availability. In particular, controllers are grouped into the clusters. In each cluster, there is a master controller plus a hot-standby instance to handle the failure scenarios. It is organized in a hierarchical manner so that the scalability is guaranteed.

3.5.2 Reliability

In evaluating a network design, the network resilience is an important factor, as a failure of a few milliseconds may easily result in terabyte data losses on high-speed links. In traditional networks, where both control and data packets are transmitted on the same link, the control and data information are equally affected when a failure happens. The existing work on the network resilience analysis have therefore assumed an in-band control model, meaning that the control plane and data plane have the same resilience properties. However, this model is not applicable to split-architecture networks. On one hand, the control packets in split-architecture networks can be transmitted on different paths from the data packet (or even on a separate network). Therefore, the reliability of the control plane in these networks is no longer linked with that of the forwarding plane. On the other hand, disconnection between controller and the forwarding planes in the split architecture could disable

Fig. 3.4 Example of
controller and switch
connection

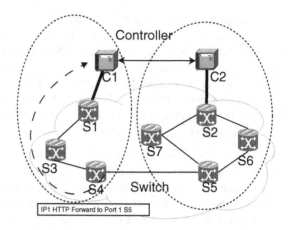

the forwarding plane: When a switch is disconnected from its control plane, it cannot receive any instructions on how to forward new flows, and becomes practically offline.

In the following, we illustrate the reliability of SDN in an example in Fig. 3.4, which consists of 7 OpenFlow switches and 2 controllers. For simplicity of illustration, we assume fixed binding between controller and switches, which is the shortest path between the switch and its closest controller. Another assumption is the static binding between controller and the switch, e.g., C1 is the assigned controller for S3. S3 can only be controlled by C1 even if it is also reachable by C2. In this example we assume there is a separate link between two controllers C1 and C2 to exchange the network states between them. Each controller uses the same infrastructure (network) to reach the OpenFlow switches. For instance, S7 goes through S3 and S1 to reach the controller C1, marked as dotted line. We also assume fixed routing has been set up. The subscripts denote the flow entries in each switch. An entry on S4 is programmed by C1 to match any HTTP flow from IP1 and forward to port 1 connected to S7.

If the link between S4 and S5 fails, connections between any of switches S1; S3; S4 to any of switches S2; S5; S6; S7 would be interrupted. If the link between S1 and controller C1 fails, then until a backup path is built and used, S1 will lose its connection to its controller. Assuming that in this case the switch invalidates all its entries, and then S1 cannot reach any other switch in the network, until it reconnects to its controller. This is like S1 itself is failed for a period of time.

3.5.3 A Special Study: Controller to Switch Connectivity

In the following, we illustrate the reliability issue using a specific problem. We focus on the controller placement problem given the distribution of forwarding plane switches. We consider that control platform consists of a set of commodity servers connecting to one or more of the switches. Therefore, the control plane and data plane are in the same network domain.

The connectivity to the controller is extremely important for the OpenFlow network reliability. We define the reliability of controller to data plane as the average likelihood of loss of connectivity between the controller and any of the OpenFlow switches.

In the following, we discuss three aspects of the connectivity between controller and the switches.

Routing Between Controller and Switches

For a give controller location, the controller can construct any desired routing tree, e.g., a routing tree that maximizes the protection of the network against component failures or a routing tree that optimizes the performance based on any desired metrics. One of the popular routing method is the shortest path routing constructed by intra-domain routing protocols such as open shortest path first (OSPF). The main problem with the shortest path routing policy is that it does not consider the network resilience (protection) factor. To maximize the routing, one can develop an algorithm with the objective of constructing a shortest path tree. Among all possible shortest path trees, we can find the one that results in best resilience compared to other shortest path trees.

Deploying Multiple Controllers

Next, we consider the problem of deploying multiple controllers in the network. The problem can be formulated as following. Given a network graph, with node representing network's switches, and edge representing network's links (which are assumed to be bidirectional). The objective is to pick a subset of the nodes, among all candidate nodes, and co-locate controllers with switches in these nodes so that the total failure likelihood is minimized. Once these nodes are selected, a solution to assign switches to controllers, is also needed to achieve maximum resilience. The problem can be solved as a graph partitioning or clustering problem. The details are described in [91]. It is shown that the choices of controller locations do have significant impact on the entire SDN reliability.

3.6 Summary

SDN architecture introduces a separation between the control and forwarding components of the network. Among the use cases of such architecture are the access/aggregation domain of carrier-grade networks, mobile backhaul, data center networks and cloud infrastructure, all of which are among the main building blocks of todays network infrastructure. Therefore, proper design, management and performance optimization of these networks are of great importance.

In this chapter, we first provided an overview of the design principles and building blocks of the SDN architecture. We then focused on two use case scenarios, i.e., access/aggregation mobile backhaul networks and inter-data center WAN. We further discussed a few practical issues in deploying SDN in the production network, including the issues of scalability and reliability. As a case study, we explained the design of controller to switch conductivities in details.

Chapter 4
Virtualizing the Network Services: Network Function Virtualization

Network function virtualization (NFV) is rapidly emerging as the de facto approach operators will use to deploy their networks. NFV leverages on cloud computing principles to change the way NFs like gateways and middleboxes are offered. As opposed to today's tightly coupling between the NF software and dedicated hardware, NFV concept requires the virtualization of NFs allowing them to run on generalized hardware.

NFV is becoming the norm by which, operators will deploy their networks. NFV leverages on cloud computing principles with the aim to transform the way NFs are implemented and deployed (e.g., firewalls, content filters, IDS, DPI, NAT, BRAS, PE routers, etc.). NFV calls for the virtualization of legacy NFs previously offered by specialized equipment/middleboxes. vNFs can run on generalized hardware and be consolidated in the operator's data centers/clouds. NFV advertises cost saving benefits and a lower CAPEX through commoditization of hardware and cloud-based pay-as-you-grow models. Furthermore, NFV reduces Opex through the centralized and unified management of NFs. Time to market of new services is also to improve, allowing operators to innovate and monetize their network at a faster pace. In this chapter, we introduce the concept of NFV, its architecture, use cases, challenges, and research problems.

4.1 NFV Architecture

To meet the increasing traffic demands while maintaining or improving ARPU, network operators are constantly seeking new ways to reduce their OPEX and CAPEX. To this end, the concept of NFV was initiated within the ETSI consortium [92]. NFV allows legacy NFs offered by specialized equipment to run in software on generic hardware. Therefore, NFV makes it possible to deploy vNFs in high-performance

© Springer International Publishing AG 2017
M. Vaezi and Y. Zhang, *Cloud Mobile Networks*, Wireless Networks,
DOI 10.1007/978-3-319-54496-0_4

commodity servers in an operator's data center, with great flexibility to spin on/off
the vNFs on demand. In addition, by decoupling the NF software from the hardware,
NFV will facilitate a faster pace for innovations and result in shorter time to market
for new services.

The recent advances in software engineering and high-performance commodity
servers motivated virtualization of NF in DC/cloud. This technology is coined as
NFV and has gained an increased popularity by network operators. NVF allows the
NFs traditionally delivered on proprietary and application-specific hardware to be
realized in software, and to run on generic commercial off the shelf servers. NFV
sets out to achieve high resource utilization and shorter service development cycles,
with a great potential for CAPEX/OPEX savings and ARPU respectively.

NFV covers a wide spectrum of middleboxes such as firewalls, DPI, IDS, NAT,
WAN accelerators, etc. It also covers a variety of network nodes such as broadband remote access servers data network gateways (S-GW/P-GW), MME, home
subscriber server (HSS), and virtual IP multimedia subsystem (vIMS) for virtual
evolved packer core (vEPC). These are the critical devices in the mobile broadband
and cellular networks. Figure 4.1 shows a set of vNFs.

NFV leverages on cloud computing principles to change the way NFs like gateways and middleboxes are offered. As opposed to today's tightly coupling between
the NF software and dedicated hardware, NFV concept requires the virtualization
of NFs allowing them to run on generalized hardware. In the past 5 years, NFV
has grown from research proposal to real deployment. There have been various
industry activities to make it prosper. The European Telecommunications Standards Institute (ETSI) consortium has published the NFV architecture documents
and has formed various working group to study the design of different aspects of
NFV. Figure 4.2 shows the ETSI NFV reference architecture [92]. It includes the
following components.

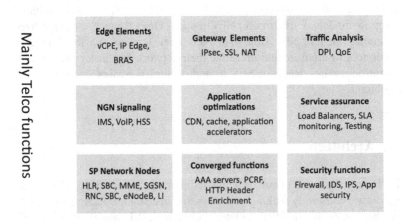

Fig. 4.1 Virtualized network functions landscape

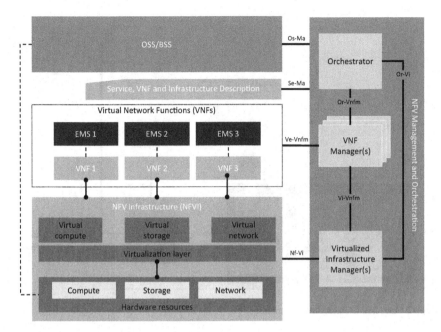

Fig. 4.2 ETSI NFV reference architecture

- NFV Orchestrator: On the one hand, it handles the network wide orchestration and management of NFV (infrastructure and software) resources. On the other hand, it is used to realize the NFV service topology on NFVI.
- VNF Manager(s): It is responsible for vNF lifecycle management, including vNF creation, resource allocation, migration, and termination.
- Virtualized Infrastructure Manager(s): it is responsible for controlling and managing the computing, storage and network resources, as well as their virtualization.
- OSS: Similar to OSS in traditional cellular network infrastructure, the main role of OSS in NFV is supporting the comprehensive management and operations functions.
- EMS: It performs the typical FCAPS functionality for a vNF. The current element manager functionality needs to be adapted for the virtualized environment.

NFV is tightly related to various new technologies, including the network virtualization, SDN, and cloud computing. In the next five years, NFV will continue to grow and will have a wide deployment in Telecom networks. For example, AT&T has announced Domain 2.0 project which will fundamentally transform their infrastructure using NFV and SDN technologies in the next five years. NFV will also have broad applicability in large enterprise networks to reduce their CPEX and OPEX.

The open source community OPNFV has been growing and pushing the open source development of the NFV infrastructure. One open source NFV activity is called OPNFV [84]. Figure 4.3 shows the OPNFV architecture. It provides the

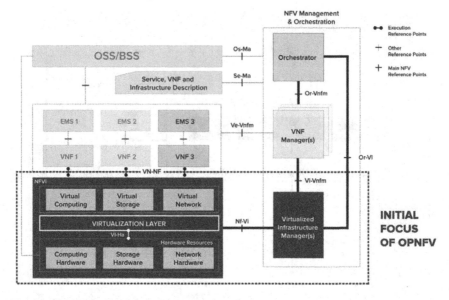

Fig. 4.3 OPNFV architecture framework

best carrier-grade NFV infrastructure (NFVI), virtualized infrastructure management (VIM) and APIs to other NFV elements. It forms the basic infrastructure required for vNFs and management and network orchestration (MANO) components.

4.2   NFV Use Cases

Virtualizing NFs could potentially offer many benefits including, but not limited to:

- Reduced equipment costs and reduced power consumption through consolidating equipment and exploiting the economies of scale of the IT industry.
- Increased speed of Time to Market by minimizing the typical network operator cycle of innovation. Economies of scale required covering investments in hardware-based functionalities are no longer applicable for software-based development, making feasible other modes of feature evolution. NFV should enable network operators to significantly reduce the maturation cycle.
- Availability of network appliance multi-version and multi-tenancy, which allows the use of a single platform for different applications, users, and tenants. This allows network operators to share resources across services and across different customer bases.
- Targeted service introduction based on geography or customer sets is possible. Services can be rapidly scaled up/down as required.

- Enables a wide variety of ecosystems and encourages openness. It opens the virtual appliance market to pure software entrants, small players and academia, encouraging more innovation to bring new services and new revenue streams quickly at much lower risk.

4.3 NFV Challenges

Elasticity: Building on top of the virtualization technology, an NFV platform should be able to leverage the benefit of running instances in the cloud: multiplexing and dynamical scaling. For multiplexing, it allows the same NF instance to serve multiple end users in order to maximize the resource utilization of the NF. On the other hand, for dynamical scaling, when the demand changes, the network operators should be able to dynamically increase/decrease the number and/or size of each NF type to accommodate the changing demands. This, in turn, will allow the Telecom Service Providers to offer their customers with the "pay as you grow" business models and avoid provisioning for peak traffic. It should support subscriber-based, application-based, device-based, and operator-specific policies simultaneously. Moreover, adding or removing new NFs should be easily manageable by the network operator, without requiring the physical presence of technicians on the site or having the enterprise customers involved. It should also be possible to accurately monitor and reroute network traffic as defined by policy.

Openness: Aligned with the Open NFV strategy in the HP NFV Business unit, the NFV framework should be capable of accommodating a wide range of NFs in a nonintrusive fashion. It should support open-source based and standard solutions as much as possible, meaning that the NFs should be implemented, deployed and managed by operators, enterprises or third-party software vendors.

Carrier-grade properties: the Telecom Service Providers have high requirement on the performance, scalability, fault tolerance, and security on the solutions.

- Efficiency: the NFV platform should provide the tight NF SLAs on performance or availability, identical to the SLAs offered with dedicated services. For example, the SLA may specify the average delay, bandwidth, and the availability for all the services provided to one customer. To support the SLA compliance, the platform should closely monitor the performance for each customer and dynamically adapt the resources to meet the SLAs.
- Scalability: the platform should support a large number of vNFs and scale as the number of subscribers/applications/ traffic volume grow. The ability to offer a per-customer selection of NFs could potentially lead to the creation of new offerings and hence new ways for operators to monetize their networks.
- Reliability: the platform should abide by NFV reliability requirements. Service availability, as defined by NFV, refers to the end-to-end service availability which includes all the elements in the end-to-end service (vNFs and infrastructure components).

4.4   Service Function Chaining

Service chaining is required if the traffic needs to go through more than one inline service. Moreover, if more than one chain of services is possible, then the operator needs to configure the networking infrastructure to direct the right traffic through the right inline service path. In this invention, traffic steering refers to leading the traffic through the right inline service path.

4.4.1   Openflow-Based SFC Solution

Recently, there have been some efforts on how to steer traffic to provide inline service chaining. These mechanisms are designed to explicitly insert the inline services on the path between the endpoints, or explicitly route traffic through different middleboxes according to the policies. Simple [93] proposes a SDN framework to route traffic through a flexible set of service chains while balancing the load across Network Functions. FlowTags [94] can support dynamic service chaining. In this section, we summarize various solutions below.

- Single box running multiple services: This approach consolidates all inline services into a single box and hence avoids the need for dealing with inline service chaining configuration of the middleboxes. The operator adds new services by adding additional service cards to its router or gateway. This approach cannot satisfy the openness requirement as it is hard to integrate existing third-party service appliances. This solution also suffers from a scalability issue as the number of services and the aggregated bandwidth is limited by the router's capacity. The number of slots in chassis is also limited.
- Statically configured service chains: The second approach is to configure one or more static service chains where each service is configured to send traffic to the next service in its chain. A router classifies incoming traffic and forward it to services at the head of each chain based on the result of the classification. However, this approach does not support the definition of policies in a centralized manner and instead requires that each service to be configured to classify and steer traffic to the appropriate next service. This approach requires a large amount of service specific configuration and is error-prone. It lacks flexibility as it does not support the steering of traffic on a per subscriber basis and limits the different service chains that can be configured. Getting around these limitations would require additional configuration on each service to classify and steer traffic.
- Policy-based routing: A third approach is to use a router using policy-based routing and for each service to be configured to return traffic back to the router after processing it. The router classifies traffic after each service hop and forward it to the appropriate service based on the result of the classification. However, it suffers from scalability issues as traffic is forced through the router after every service.

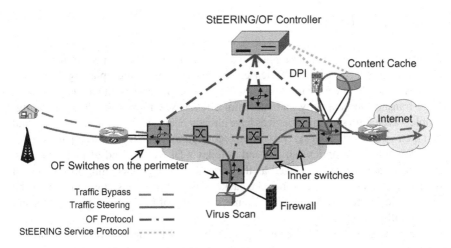

Fig. 4.4 SDN service function chaining architecture

The router must be able to handle N times the incoming traffic line rate to support a chain with N 1 services.

- Policy-aware switching layer: Recently, it is proposed to use a policy-aware switching layer for data centers which explicitly forwards traffic through different sequences of middleboxes. Each policy needs to be translated into a set of low-level forwarding rules on all the relevant switches. Often SDN is used for programming these policies. We will elaborate more next.

The components of StEERING [95] are shown in Fig. 4.4. Our system uses a logically centralized OpenFlow-based controller to manage both switches and middleboxes. The solid red line and the dotted green line in Fig. 4.4 show two different service paths. In our design, the service paths are unidirectional, that is, different service paths are specified for upstream and downstream traffic. The red line in this figure shows a service path for the upstream traffic through Virus Scan, DPI, and Content Cache. The green line shows a service path that bypasses all the services StEERING architecture uses two different types of switches. The Perimeter OF Switches are placed on the perimeter of the service delivery network. These switches will classify the incoming traffic and steer it towards the next service in the chain. These are the switches to which services or gateway nodes are connected. The Inner Switches will forward the traffic using efficient L2 switching. These switches are only connected to other switches. These switches may or may not be OF switches.

Traffic steering is a two-step process. The first step classifies incoming packets and assigns them a service path based on predefined subscriber and application. The second step forwards packets to a next service based on its current position along its assigned service path. This two-step traffic steering process only needs to be performed once between any two border routers, regardless of the number of switches that connects them.

4.4.2 Optical SFC

Optical communications have already enabled terabits level high-speed transmission. One promising technology is the dense wavelength-division multiplexing (DWDM). It allows a single fiber to carry tens of wavelength channels simultaneously, offering huge transmission capacity and spectrum efficiency. On the other hand, reconfigurable wavelength switching devices have already been widely deployed in long-haul and metro transport networks, providing reconfiguration on layer-0 lightpath topologies. We argue that optics can be used in today's DCs to the end-of-rack switches, the top-of-rack switches, as well as the servers. Although switching in the optical domain may have less agility than the packet-based approaches, it is suitable for the dynamic level required by service chains consisting of high-capacity core NFs and use of traffic aggregation.

We propose that optical technology can be used to support traffic steering. In the following, we present a packet/optical hybrid DC architecture, which enables steering large aggregated flows in an optical steering domain. Figure 4.5 illustrates an overview of the proposed architecture. The centralized OSS/BSS module interfaces with an SDN controller and a cloud/NFV manager.

The cloud manager is responsible for cloud resource allocation and automating the provisioning of virtual machines (VMs) for vNFs. It also includes an NFV management module that handles instantiation of the required vNFs while ensuring correctness of configuration.

The SDN controller can be part of the cloud management subsystem or a separate entity. The SDN controller and cloud/NFV manager perform resource provisioning. On the southbound interface, the SDN controller uses optical circuit switching to

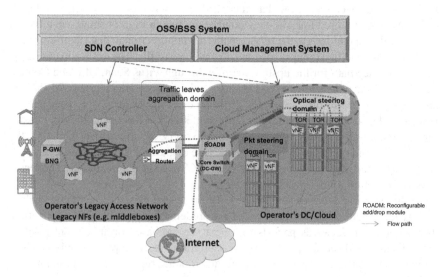

Fig. 4.5 Optical service function chaining framework

control the network elements. This interface can be realized using OpenFlow v. 1.4, which has an extension for optical circuit configuration.

To perform service chaining, the OSS/BSS module needs to request vNFs and network resources and the policies for resource allocation. The SDN controller performs network resource allocation by relying on a path computation entity that could be integrated with the SDN controller.

The data center contains both an optical steering domain and a packet steering domain. The optical steering domain conducts traffic steering for large aggregated flows. The entry point is an ROADM, which either forwards a wavelength flow to the optical domain or sends it to the packet domain for fine-grained processing. After a wavelength flow has gone through the needed vNFs, it is steered back to the ROADM. The flow is controlled to route either back to the packet domain for fine-grained processing, or forward to the optical domain for high-capacity optical processing.

4.4.3 Virtualized Network Function Placement

The VNF placement problem is important to the performance of the entire NFV system. Formally, the problem can be defined as selecting the locations for vNF instances in a NFV infrastructure in order to accommodate the traffic for a given set of service chain requests. Network traffic for a given service chain must visit the sequence of vNFs in the defined order. In the VNF Placement problem, one must place multiple instances of each VNF on servers, and choose the routes for each service chain. The goal is for the physical network to accommodate the traffic for all service chains. If the flows can not all be accommodated, the highest priority service chains should be placed. The network may have heterogeneous server types, so that one server type may be more efficient at running one type of vNF than others. VNF Placement is a challenging combinatorial problem, which is known to be NP-hard.

There are multiple objectives to consider with vNF placement. One objective may be to minimize operating costs and leave open servers for future VNF needs. To achieve that, one may want to host VNFs on as few servers as possible. Another objective could be to ensure low network latency for his customers. These two objectives cannot be satisfied simultaneously because the former needs to concentrate traffic in the network, but the latter requires spreading traffic out to avoid network congestion.

Existing approaches to VNF placement often try to minimize the number of servers used by the vNFs [96, 97]. They model network latency as a known fixed delay, and add constraints on the overall latency for each service chain. Mehraghdam et al. [98] is the first work that considers multiple objectives, including the minimizing number of servers used, maximizing the total unused bandwidth on network links, and minimizing the total latency across all service chains. Bari et al. [99] also considers multiple objectives such as VNF deployment costs, operating costs, penalties for service-level agreement violations, and resource fragmentation costs. However, both work models the latency as a fixed value.

We propose a different approach below. Our goal is to allow operators to select their operating point for trading-off resource usage and end-to-end SFC latency. It includes a mixed integer programming (MIP) model that explicitly captures the effect of network traffic on latency while maintaining a linear model: its objective is to minimize the maximum utilization over resources in the network. Minimizing the worst-case utilization avoids the situation in which a small number of congested resources induce outsized delays on network traffic. The optimization method produces a set of solutions to the VNF placement problem, each representing a different trade-off between network performance and resource usage. The approach is based on a MIP formulation of the VNF placement problem. Here we assume that there are multiple types of servers, each with different processing rates when assigned to host different types of vNFs.

4.5   Summary

In this chapter, we introduced the NFV including its architecture, its use cases in telecom network, and challenges in virtualizing the NFs. To further illustrate the challenges, explained the state-of-the-art solution, presented a few research problems, and sketch their solutions, including the service function chaining, optical service function chaining, and VNF placement.

Chapter 5
SDN/NFV Telco Case Studies

In this chapter, we introduce two case studies of SDN and NFV in the mobile core networks, which is a foundation of the next-generation mobile core network (5G).

5.1 Packet Core

The 3G packet core (PC) network consists of three interacting domains: core network (CN), 3G PC terrestrial radio access network (UTRAN), and user equipment (UE). The main function of the core network is to provide switching, routing, and transit for user traffic. Core network also contains the databases and network management functions. It is the common packet core network for GSM/GPRS, WCDMA/HSPA, and non-3GPP mobile networks. The packet core system is used for transmitting IP packets.

The core network is divided into circuit-switched and packet-switched domains. Some of the circuit-switched elements are mobile switching center (MSC), visitor location register (VLR), and gateway MSC. Packet-switched elements are SGSN and GGSN. Some network elements, like EIR, HLR, VLR, and AUC, are shared by both domains.

The architecture of the core network may change when new services and features are introduced. Number portability database (NPDB) will be used to enable user to change the network while keeping their old phone number. Gateway location register (GLR) may be used to optimize the subscriber handling between network boundaries. The primary functions of the packet core with respect to mobile wireless networking are mobility management and QoS. These functions are not typically provided in a fixed broadband network but they are crucial for wireless networks. Mobility management is necessary to ensure packet network connectivity when a wireless terminal moves from one base station to another. QoS is necessary because, unlike fixed networks, the wireless link is severely constrained in how much bandwidth it can provide to the terminal, so the bandwidth needs to be managed more tightly than in fixed networks in order to provide the user with an acceptable quality of service.

© Springer International Publishing AG 2017
M. Vaezi and Y. Zhang, *Cloud Mobile Networks*, Wireless Networks,
DOI 10.1007/978-3-319-54496-0_5

The signaling for implementing the mobility management and QoS functions is provided by the GPRS tunneling protocol (GTP). GTP has two components:

- GTP-C a control plane protocol that supports establishment of tunnels for mobility management and bearers for QoS management that matches wired backhaul and packet core QoS to radio link QoS
- GTP-U a data plane protocol used for implementing tunnels between network elements that act as routers. There are two versions of GTP-C protocol, i.e., GTP version 1 (GTPv1-C and GTPv1-U) and GTP version 2-C (designed for LTE). In this invention, we focus on GTPv1 and the 3G PC-based system.

Network services are considered end-to-end, this means from a terminal equipment to another. An end-to-end service may have a certain QoS which is provided for the user of a network service. It is the user that decides whether he is satisfied with the provided QoS or not. To realize a certain network QoS service with clearly defined characteristics and functionality is to be set up from the source to the destination of a service. In addition to the QoS parameters, each bearer has an associated GTP tunnel. A GTP tunnel consists of the IP address of the tunnel endpoint nodes (radio base station, SGSN, and GGSN), a source and destination UDP port, and a tunnel endpoint identifier (TEID). GTP tunnels are unidirectional, so each bearer is associated with two TEIDs, one for the uplink and one for the downlink tunnel. One set of GTP tunnels (uplink and downlink) extends between the radio base station and the SGSN and one set extends between the SGSN and the GGSN. The UDP destination port number for GTP-U is 2152 while the destination port number for GTP-C is 2123. The source port number is dynamically allocated by the sending node.

5.1.1 Existing Solutions Problems

The 3GPP standards do not specify how the packet core should be implemented; they only specify the network entities (SGSN, etc.), the functions each network entity should provide, and the interfaces and protocols by which the network entities communicate. Most implementations of the packet core use servers or pools of servers dedicated to a specific network entity. For example, a pool of servers may be set up to host SGSNs. When additional signaling demand requires extra capacity, an additional SGSN instance is started on the server pool, but when demand is low for the SGSN and high for, for example, the HSS, the HSS servers will be busy while the SGSN servers may remain underutilized. In addition, server pools that are underutilized will still consume power and require cooling even though they are essentially not doing any useful work. This results in an additional expense to the operator.

An increasing trend in mobile networks is for managed services companies to build and run mobile operator networks, while the mobile operator itself handles marketing, billing, and customer relations. Mobile operator managed services companies may have contracts with multiple competing operators in a single geographic region. A mobile operator has a reasonable expectation that the signaling and data traffic for their network is kept private and that isolation is maintained between the traffic for their network and for that of their competitors, even though their network and their competitors' networks may be managed by the same managed services company. The implementation technology described above requires the managed services company to maintain a completely separate server pool and physical signaling network for each mobile operator under contract. The result is that there is a large duplication of underutilized server capacity, in addition to additional power and cooling requirements, between the operators.

The packet core architecture also contains little flexibility for specialized treatment of user flows. Though the architecture does provide support for QoS, other sorts of treatment involving middleboxes, for example, specialized deep packet inspection or interaction with local data caching and processing resources such as might be required for transcoding or augmented reality applications, is difficult to support with the current PC architecture. Almost all such applications require the packet flows to exit through the GGSN, thereby being de-tunneled from GTP, and be processed within the wired network.

5.1.2   Virtualization and Cloud Assisted PC

The basic concept of bringing virtualization and cloud to PC is to split the control plane and the data plane for the PC network entities and to implement the control plane by deploying the EPC control plane entities in a cloud computing facility, while the data plane is implemented by a distributed collection of OpenFlow switches. The OpenFlow protocol is used to connect the two, with enhancements to support GTP routing. While the PC already has a split between the data and control plane, in the sense that the SGSN and GGSN are pure control plane, e.g., HLR, HSS, and AuC, the EPC architecture assumes a standard routed IP network for transport on top of which the mobile network entities and protocols are implemented.

The split proposed in this document is instead at the level of IP routing and MAC switching. Instead of using L2 routing and L3 internal gateway protocols to distribute IP routing and managing Ethernet and IP routing as a collection of distributed control entities, this document proposes centralizing L2 and L3 routing management in a cloud facility and controlling the routing from the cloud using OpenFlow. The standard 3G PC control plane entities, SGSN,GGSN, HSS, HLR, AuC, VLR, EIR, SMS-IWMSC, SMS-GMSC, and SLF are deployed in the cloud. The data plane consists of standard OpenFlow switches with enhancements as needed for routing GTP packets, rather than IP routers and Ethernet switches. At a minimum, the data plane traversing through the SGSN and GGSN and the packet routing part of the NodeB in the E-UTRAN require OpenFlow enhancements for GTP routing. Additional enhancements for GTP routing may be needed on other switches within the 3G PC depending on how much fine-grained control over the routing an operator requires.

The packet core control plane parts of the gateways for GTP-C communications, i.e., the parts that handle GTP signaling, are implemented in the cloud as part of the OpenFlow controller. The control plane entities and the OpenFlow controller are packaged as VMs. The API between the OpenFlow control and the control plane entities is a remote procedure call (R3G PC) interface. This implementation technology is favorable for scalable management of the control plane entities within the cloud, since it allows execution of the control plane entities and the controller to be managed separately according to demand. The cloud manager monitors the CPU utilization of the 3G PC control plane entities and the control plane traffic between the PC control plane entities within the cloud. It also monitors the control plane traffic between the UEs and NodeBs, which do not have control plane entities in the cloud and the PC control plane entities. If the 3G PC control plane entities begin to exhibit signs of overloading, like utilizing too much CPU time or queuing up too much traffic, the overloaded control plane entity requests that the cloud manager starts up a new VM to handle the load. The cloud manager also provides reliability and failover by restarting a VM for a particular control plane function if any of the PC control plane entities should crash, collecting diagnostic data, saving any core files of the failed PC control plane entity, and informing the system administrators that a failure occurred. The control plane entities maintain the same protocol interface between themselves as in the standard 3GPP 3G PC architecture.

The OpenFlow control plane, shown here as a gray-dotted line, manages the routing and switching configuration in the network. The OpenFlow control plane connects the SGSNs, the standard OpenFlow switches, and the GGSN to the Open-Flow controller in the cloud. The physical implementation of the OpenFlow control plane may be as a completely separate physical network, or it may be a virtual network running on the same physical network as the data plane, implemented with a prioritized VLAN or with an MPLS label-switched path or even with a GRE or other IP tunnel. The OpenFlow control plane can in principle use the same physical control plane paths as the GTP-C and other mobile network signaling. The SGSN-Ds and the GGSN-Ds act as OpenFlow GTP-extended gateways, encapsulating and decapsulating packets using the OpenFlow GTP extensions.

The NodeBs have no control plane entities in the cloud because the RAN signaling required between the RNC and the NodeB includes radio parameters, and not just IP routing parameters. Therefore, there is no OpenFlow control plane connection between the OpenFlow controller in the cloud and the NodeBs. The NodeBs can, however, act as OpenFlow GTP-extended gateways by implementing a local control to data plane connection using OpenFlow. This allows the packet switching side of the NodeBs to utilize the same OpenFlow GTP switching extensions as the packet gateways.

The operation of the PC cloud works as follows. The UE, NodeB, SGSN, and GGSN signal to the HLR, HSS, AuC, and SMS-GMSC using the standard EPC protocols, to establish, modify, and delete GTP tunnels. This signaling triggers procedure calls with the OpenFlow controller to modify the routing in the EPC as requested. The OpenFlow controller configures the standard OpenFlow switches, the Openflow SGSN, and GGSN module with flow rules and actions to enable the routing requested

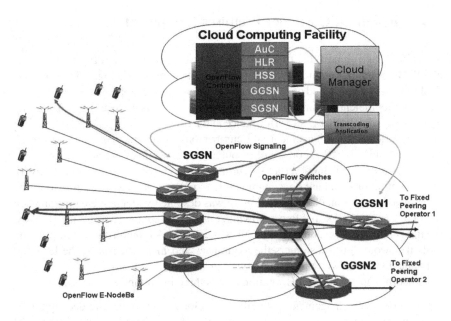

Fig. 5.1 Virtualized PC and SDN routing

by the control plane entities. Details of this configuration will be described in the next section.

Figure 5.1 illustrates how PC peering and differential routing for specialized service treatment are implemented. These flow rules steer GTP flows to particular locations. The operator, in this case, peers its PC with two other fixed operators. Routing through each peering point is handled by the respective GGSN1-D and GGSN2-D. The blue arrow shows traffic from a UE that needs to be routed to either one or another peering operator. The flow rules and actions to distinguish which peering point the traffic should traverse are installed in the OpenFlow switches and gateways by the OpenFlow controller. The OpenFlow controller calculates these flow rules and actions based on the routing tables it maintains for outside traffic, and the source and destination of the packets, as well as by any specialized forwarding treatment required for DSCP marked packets.

The red arrow shows an example of a UE that is obtaining content from an external source. The content is originally not formulated for the UE's screen, so the OpenFlow controller has installed flow rules and actions on the GGSN1, SGSN-D, and the OpenFlow switches to route the flow through a transcoding application in the cloud. The transcoding application reformats the content so that it will fit on the UE's screen. MSC requests the specialized treatment at the time the UE sets up its session with the external content source via the IP multimedia subsystem (IMS) or another signaling protocol.

5.2 Virtualized Customer Premises Equipment

Virtualizing and cloudifying the customer premises equipment (CPE) of enterprises and SMBs is an important NFV use case. In practice, most existing vCPE PoCs and deployments are overlayed on a distributed physical network topology with relatively static and inefficient resource placement. The current focus has been on connectivity and functionality, rather than performance and agility.

The static resource placement and cumbersome deployment are a result of the need to provide high SLAs on top of resources and tools that were not planned to do so. For example, existing overlay solutions are not aware of the underlay network and its limitations, and are hence vulnerable to reduced service levels due to traffic dynamics. In addition, cloud provisioning was designed with compute optimization in mind, whereas vNFs are often bandwidth rather than compute constrained. Finally, the industry has not yet come up with NFV specific SLA monitoring and verification tools that would give customers the assurances and make them trust the lean and dynamic distributed systems with carrying production size loads.

Cloudification of the vCPE solution brings three important advantages.

- Cost-effective management and agility: By cloudifying vCPE, we decouple provisioning and vNF onboarding which is complex and slow today, due to the local provisioning of the physical enterprise site. This semi-manual provisioning may be frequent and dependent on many access network factors, including changing demands of the enterprise. In the cloudified solution, the vCPE capacity of all the enterprise sites is in the distributed carrier cloud, e.g., the provider point of presence (POP) sites.
- High performance at scale: By cloudifying vCPE we only need to orchestrate resources per POP for the average demand of the enterprise sites connected to that POP. This includes also Internet peering capacity and related NAT functions. We also do not need to explicitly orchestrate for geo-redundant vCPE capacity. This can lead to significant (some estimates are upto 4x) savings in vCPE resources.
- High availability: By cloudifying vCPE, we can have faster recovery from vCPE failures. Traffic to failed components is distributed to cloudified resources avoiding sharp hits and potential domino collapses in case of POP or rack failures.

To achieve the goals of cloudification with scalable, dynamic and efficient operations we must have the following:

- Network underlay awareness: If the demand is randomly distributed to the POP overlay edges, without the knowledge of the underlay, congestion and packet loss could occur, resulting in SLA violations. This happens in the normal course of mapping enterprise flows to vCPE resources and Internet peering, and it is true when remapping enterprise flows in case of failures.
- SLA verification: By cloudifying and dynamically mapping traffic to resources we gain savings and decoupled manageability but we now have to prove per flow that we still meet the enterprise SLA just as well as with static local provisioning. Per

flow SLA and connectivity verification can also trigger additional VM provisioning without interrupting services.

- Resource defragmentation: As we decouple orchestration and the system is running events will trigger additional allocations, extended service chains, compensation for blade, or CPU failures. But because of cloudification we can constantly run in the background proceeds that reallocate vCPEs in dense hardware configuration, gracefully ramping down fragmented vNFs till they are "garbage collected" freeing hardware for further dense orchestration.

5.2.1 Requirements

The design of our vCPE platform is based on a set of key requirements we identified by analyzing current SLA structure and management models as well as SMEs common pain points. Our approach highlights elasticity, flexibility, efficiency, scalability, reliability, and openness as critical components to support the goals of NFV.

- Elasticity: Building on top of NFV, vCPE should be able to leverage the benefit of running instances in the cloud: multiplexing and dynamical scaling. For multiplexing, it allows the same NF instance to serve multiple end users in order to maximize the resource utilization of the NF. On the other hand, for dynamical scaling, when the demand changes, the network operators should be able to dynamically increase/decrease the number and/or size of each NF type to accommodate the changing demands. This in turn will allow the enterprise customer to benefit from pay as you grow business models and avoid provisioning for peak traffic.
- Flexibility: The vCPE platform should support subscriber-based, application-based, device-based, and operator specific policies simultaneously. Moreover, adding or removing new NFs should be easily manageable by the network operator, without requiring physical presence of technicians on the site or having the enterprise customers involved. It should also be possible to accurately monitor and reroute network traffic as defined by policy. The platform should allow NFs to be implemented, deployed, and managed by operators, enterprises or third party software vendors.
- Efficiency: The vCPE should provide the tight NF service-level agreements (SLAs) on performance or availability, identical to the SLAs offered with dedicated services. For example, the SLA may specify the average delay, bandwidth, and the availability for all the services provided to one customer. To support the SLA compliance, the platform should closely monitor the performance for each customer and dynamically adapt the resources to meet the SLAs.
- Scalability: The vCPE framework should support a large number of rules and scale as the number of subscribers/applications/traffic volume grows. The ability to offer a per-customer selection of NFs could potentially lead to the creation of new offerings and hence new ways for operators to monetize their networks.

- Reliability: The vCPE framework should abide by NFV reliability requirements. Service availability as defined by NFV refers to the end-to-end service availability which includes all the elements in the end-to-end service (vNFs and infrastructure components) with the exception of the end user terminal.
- Openness: The final issue is ensuring that the vCPE framework should be capable of accommodating a wide range of NFs in a non-intrusive fashion. The vCPE should support open-source based and standard solutions as much as possible.

5.2.2 Design

In the vCPE architecture, an SME is connected to the Internet via a lightweight CPE also called SDN-based CPE. Most features typically found on a CPE, like NAT, DHCP, a firewall/security gateway, are moved to VMs in the operator's cloud. The lightweight SDN-based CPE only retains the basic connectivity service, while being programmable via SDN.

In the operator's cloud, VMs and physical resources (storage, compute) are configured via a cloud controller while network configuration is managed by an SDN controller. Finally, a moderator node/application provides a higher level of abstraction over the cloud and network controllers. Through the moderator enterprises can select their desired services and configure service chaining policies. The enterprise's IT personnel can access the moderator through a secured dashboard with a user-friendly graphical interface. Figure 5.2 shows these components and their interactions; each is further described below.

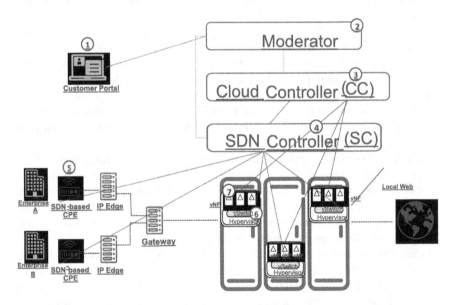

Fig. 5.2 Virtualized CPE

- The customer portal: Through the customer portal, an enterprise administrator configures and manages enterprise policies, services, and network infrastructure. Each enterprise gets its own virtual infrastructure. VMs are launched for each enterprise and are not shared between enterprises. The first step is to register the vCPE along with its name, IP block, subnets, etc. Then, through the same portal, the enterprise specifies how the traffic should be mapped and steered across the vNFs (i.e., the service chaining policies).

- The moderator presents services and selected service chains to the enterprise customer, and abstracts away most details of resource allocation for VMs and network configuration. Each enterprise has a catalog of available services. Services can be deployed, and can be chained in arbitrary order in both upstream and downstream directions.

- The cloud controller (CC) is a typical cloud controller (e.g., OpenStack) augmented with support for flow networks (i.e., flow network extensions added to the neutron). The moderator translates the list of services and their connectivity into information about VMs, vSwitches, and links that the CC can understand. The CC receives the customer's network architecture and policy specifications, akin to a fine-grained SLA. It translates the SLA into a list of requirements in terms of dedicated VMs, storage, different types of network appliances, and business applications, as well as dedicated links between those. Next, it maps a constructed virtual topology onto the network abstraction provided by the SC (SDN controller). Based on SC feedback, the CC proceeds to create and configure the customer infrastructure (i.e., instantiate VMs, virtual switches). The CC informs the SC of the placement of specific network entities, like virtual switches.

- The SDN controller (SC) is responsible for managing and provisioning the enterprise network topology, by mapping network requirements to the selected set of physical and virtual network resources (including customer's CPE). Such configuration is done using a combination of different southbound plugins, like OpenFlow, OVSDB, or NetConf. The SC is an SDN controller with a developed application for vCPE service chaining. Some of the extensions require changes to the external interfaces of the SDN controller. The main extensions for vCPE include service chaining, connectivity monitoring, location optimization, and network configuration. Service chaining provided the APIs for orchestrator to create service chaining rules per enterprise. Location optimization and connectivity monitoring provide APIs to detect network congestion and connectivity failures. This information is necessary to enforce network SLAs. Network configuration provides the orchestrator the possibility to engineer service chaining networks in many different ways. The SC will also inform the CC of optimal locations to instantiate and interconnect VMs. It also notifies of link congestion or failure, in order to trigger VM migrations and network re-configuration.

- The SDN-based CPE and vSwitch is a lightweight version of legacy CPE with most NFs stripped out. The virtual switch is a software-based OpenFlow switch such as Open vSwitch. The vNFs execute in VMs on top of a hypervisor that is hosted on a server. Multiple virtual services can share the resources of the same server. We assume that one instance of a vSwitch is included in each server for handling the communication between multiple VMs running on the servers within the data center. Both SDN-based CPE and vSwitch are programmable by SDN to support the vCPE applications including service chaining.

5.3 Summary

In this chapter, we presented the SDN and NFV's use cases in the telecom network as two case studies. In particular, we first discussed their usage scenarios and challenges in the packet core network. Next, we discussed their usage in the edge of the telecom network, in the form of customer premise equipment. Both use cases have been widely studied and are going to deployment in the real world.

Chapter 6
Radio Access Network Evolution

Today, mobile network operators have a variety of options for the means of providing more coverage, capacity, and services to their subscribers. Alongside the advancement of wireless standards to provide higher spectral efficiency, vendors have released a wide range of radio access nodes for network densification. Starting with 3G and especially LTE deployments, legacy macro BSs are replaced with *distributed* BS structures.[1] While the macrocell remains the workhorse, *small cells*, and *distributed antenna systems* (DAS) are increasingly deployed, and *mobile virtual network operators* (MVNOs) are sprouting up here and there, *cloud radio access network* (cloud RAN) is expected to be deployed in near future, and *RAN-as-a-Service* (RANaaS) is gaining traction. These are some of the trends in this space which will be covered in the current and following chapters.

6.1 Mobile Network Architecture

6.1.1 End-to-End Architecture

The overall end-to-end architecture of a carrier network is composed of three big parts: the radio access network, core network, and external network. The *radio access network* is the first component of any carrier network and provides access and shuttles the voice/data to and from the user equipment. The *core network* then connects the radio access network to the external network, e.g., the public-switched telephone network or the public Internet. LTE core network, which is also known as the evolved packet core (EPC), is responsible for the data routing, accounting, and policy management; it offers services to the customers who are interconnected by the access network.

---

[1]This can be regarded as a step toward C-RAN from the traditional macro RAN architecture.

© Springer International Publishing AG 2017
M. Vaezi and Y. Zhang, *Cloud Mobile Networks*, Wireless Networks,
DOI 10.1007/978-3-319-54496-0_6

Table 6.1 Mobile backhaul network primary locations

Generation	Base station	Controller	Backhaul interface	Backhaul aggregation
2G	BTS	BSC	Abis	TDM
3G	NodeB	RNC	Iub	ATM/IP
4G	eNodeB	eNodeB, MME, and SGW	S1	IP

The radio access network (RAN) is composed of the radio base stations (BSs), base station controllers (BSCs), and backhaul network. The main components of RAN in various generations of mobile networks are listed in Table 6.1. In 2G and 3G systems, RAN was designed in a way that traffic from all BSs was sent to a controller, namely the BSC in 2G and radio network controller (RNC) in 3G, where radio resource management, some mobility management functions, and data encryptions were carried out. In LTE, BSs (eNodeBs) have direct connectivity with each other (known as X2 traffic) and host the radio resource controller (RRC) which performs all resource assignment for each active user in its cell. The mobility management entity (MME) authenticates devices and is involved in handoffs between LTE and previous generations of technology. Also, serving gateway (SGW) acts like a router for users, passing data back and forth from the user to the network, and is responsible for handovers with neighboring eNodeBs. Finally, as another component of the RAN, the backhaul network[2] acts as the link between BSs and BSCs, toward the core.

6.1.2   Packet Flow in a Mobile Network

Among the current widely deployed 4G systems, i.e., HSPA+, WIMAX, and LTE, the latter two are pure packet-based networks without traditional voice circuit capabilities; they provide voice services via IP. Hence, it is important to know the different steps of packets transmission in a mobile network.

Assume that the user equipment (UE) is in the idle state. That is, it has already authenticated with a 4G network, but it does not have an RRC connection yet. Once the user types in a URL and hits "Go" an RRC connection is established and the UE turns into the connected state, and then into the active state when it enters the dedicated mode. The 3GPP defines two types of latency, namely, *user plane latency* and *control plane latency*. The user plane latency refers to the one-way transmit time of a data packet at the IP layer in the UE to the IP layer in the BS, or vice versa [100, 101]. The control plane latency is the time required for the UE to transit from idle state to active state.

The user and control plan latencies only define the delay in transferring packets from the device to the BS. However, the packets need to travel through the core

---

[2]Mobile backhaul network will be elaborated in Sect. 6.3.1.

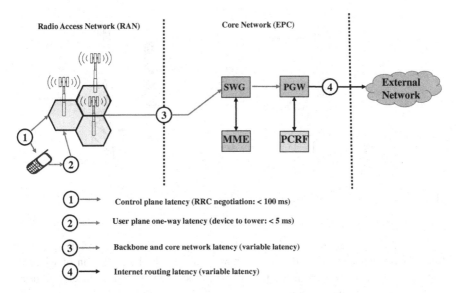

Fig. 6.1 LTE request flow latencies [5]

network and out to the public Internet. The 4G standards make no guarantees on the latency of this path. Thus, this delay varies from carrier to carrier. In many deployed 4G networks, this delay is 30–100 ms. To summarize, a user initiating a new request incurs the following latencies:

1. **Control plane latency**: Fixed, one-time latency due to RRC connection and state transitions from idle to active (<100ms) and dormant to active (<50ms).
2. **User plane latency**: Fixed latency (<5 ms) for every application packet transferred between the UE and the BS.
3. **Core network latency**: Variable (carrier-specific) latency for packet transporting from the BS to the packet gateway (practically 30–100 ms).
4. **Internet routing latency**: Variable latency between the packet gateway and the destination address on the public Internet.

The first two latencies are bounded by the standard requirements and the figures inside brackets are for the LTE system. The packet request flow in mobile networks is depicted in Fig. 6.1 [5].

6.2 Base Station Architecture

A radio base station, also known as base transceiver station (BTS), is the equipment that facilitates wireless communication between user equipment (UE) and a mobile network. As shown in Table 6.1, a BS is widely referred to as *NodeB* and *evolved*

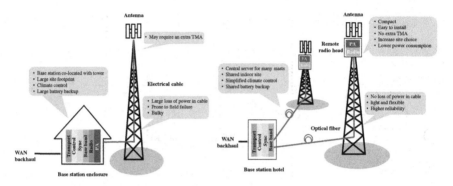

Fig. 6.2 Old-style macro BS versus distributed BS

NodeB (eNB), respectively, in 3G and LTE networks. The BSs' architecture and functionality has experienced a lot of changes alongside the evolution of wireless technologies. As it is the main component to be modified yet again in the cloud-based RAN, we study its evolution in the following.

6.2.1 Legacy BS

Traditional BSs consist of one or more racks, equipped with the baseband and radio equipment, commonly sitting at the foot of the tower, and feeding signals back and forth to passive antennas mounted on top of the tower. The antennas are connected to the racks via coaxial cables (feeders) as shown in Fig. 6.2. Such BS sites often require large, sturdy shelters and strong structural support which make site acquisition difficult and costly; they also consume a large amount of energy. Additionally, with this architecture, the signal power transmitted by the BS racks to the antennas typically encounters a loss of approximately 3 dB. Distributed BS architecture allows operators to address these challenges.

6.2.2 Architecture of Distributed BS

In a *distributed base station* the RF filter and power amplifier of a BS are installed next to the antenna on a tower or rooftop rather than at the tower bottom where traditional macro BSs are typically located. This component of the distributed BS is then called the *remote radio head* (RRH). It helps eliminate the 3 dB signal power loss since the power amplifiers residing in close proximity to the antennas. Distributed BSs have another component that contains the digital assets; it is called the *baseband unit* (BBU). The connection link down the tower (the link between RRH and BBU) is now a digital fiber optic link, such as CPRI or OBSAI [102]. Starting with 3G networks, this separate architect has been adopted in access networks.

With this modular design and by allowing the main components of a BS to be installed separately, distributed BSs use available real estate more efficiently and reduce power consumption while providing capacity comparable to the conventional macro solutions and better coverage and network performance due to the smaller loss. Figure 6.2 compares the old-style macro BS with the distributed BS.

In addition to greater deployment flexibility resulted from the fact that BBU and RRH can be separated up to several miles, distributed BS makes site acquisition easier and rental cost lower due to their smaller footprint. Besides, they consume less amount of power, up to 50% as reported in [103], which can make renewable energy applicable. Note that the RRH requires a quite compact power supply and consumes much less power, since it can work in natural heat dissipation mode. Additionally, as we will see in more detail in Chap. 7, several BBUs could be pooled at central sites making intercell communication possible, while RRHs are distributed in different cells. Such an architecture is especially favorable for LTE-Advanced (LTE-A) as features requiring coordination among neighboring cells can be easily realized at central processing site where BBUs are pooled. Several features and benefits of distributed BSs are listed in Table 6.2.

On account of the lower cost and greater deployment flexibility they offer, RRHs are currently being deployed not just for new technologies (e.g., LTE and LTE-A) but also in new and replacement infrastructure for older technologies (be it 3G, 2.5G, or 2G). However, some operators are still adopting traditional architecture as the radio equipment is easily accessible from technical personnel for maintenance and service. Note that, mounting RRH on top of the tower mandates a tower climb for troubleshooting, while in old-style BSs radio and BBU units were co-located in the hut at the ground level.

Table 6.2 Distributed BS: features and benefits

Feature	Benefit
BBU and RRH can be spaced miles apart	• Higher degree of deployment flexibility
Reduced space (footprint)	• Lower rental costs
	• Easier site acquisition
Lightweight RRH	• Easier installation
	• No need for feeders
Better coverage than old-style macro sites when deployed in tower-top (no feeder loss)	• Reduced total number of sites
	• No need for TMAs
Integrated maintenance and administration when BBUs pooled	• Reduced OPEX
Reduced power consumption (RRHs work in natural heat dissipation mode)	• Environment friendly
	• Reduced OPEX

RRH might be replaced by active antenna system (AAS), as another step in the evolution of radio networks. Active antennas integrate the radio into the (passive) antenna and distribute the radio functionality across the antenna elements. This means the RRH is integrated into the antenna. Even smaller than conventional passive antennas, the AAS offers better performance and brings the following benefits to operators:

- **Lower CAPEX and OPEX**: Integrated antenna array eliminates feeders, jumpers, and connectors to reduce RF power losses. Hence, it increases energy efficiency and thereby reduces electricity costs. The AAS also reduces installation and site costs as it is easier to install the AAS for its fewer components and eliminated tower mounted amplifier (TMA). More compact than traditional passive antenna plus RRH, the AAS could help reduce site rental costs. By increasing reliability through incorporating redundant subsystems, the AAS can reduce OPEX too.
- **Higher capacity and coverage**: Advanced features like vertical beamforming,[3] independent Rx/Tx electrical tilting per frequency and MIMO technology improve coverage and capacity [104].

6.3   Mobile Fronthaul and Backhaul

In a mobile network, the backhaul network spans the connection from the access nodes at the edge of the network through the aggregation network and then handing off the signals to the controller, and possibly to the core network. Put simply, mobile backhaul connects the BSs to the BSCs [6], with the QoS requirements of different applications. This is depicted in Fig. 6.3.

6.3.1   Backhaul

Acting as the link between the edge of a network (access nodes) and the mobile core, mobile backhaul plays an important role in mobile networks. It enables to transport mobile data from the end user to mobile networks, traditional telephone networks, and the Internet. Mobile backhaul includes *backhaul access* and *transportation/aggregation*, as illustrated in Figs. 6.3 and 6.4. Today, a variety of access types such as fiber, copper, and microwave are used for backhaul access. Meanwhile, new backhaul topologies are required to support the shift toward HetNets and C-RAN and support multiple RAT including 3G, LTE, and Wi-Fi. The aggregation network supports a large number of end users, and hence it should be protected from

---

[3]The AAS provides electrical beam control both in the horizontal and vertical plans. This offers several spatial processing techniques such as separating tilt based on frequency, RAT, service, etc., as illustrated in [104, Fig. 1].

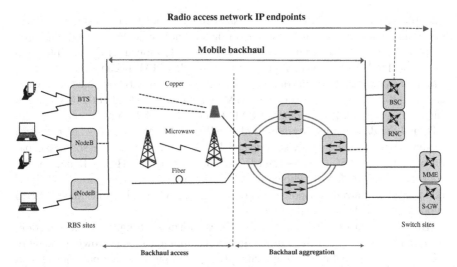

Fig. 6.3 Mobile backhaul in 2G/3G/4G networks [6]

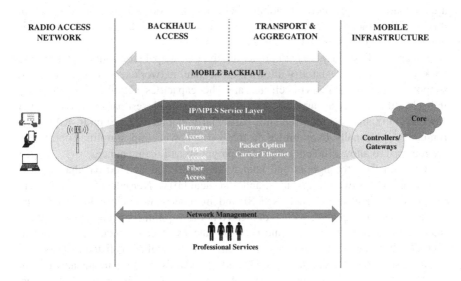

Fig. 6.4 Mobile backhaul for HetNets [7]

link and node failures. This is usually achieved through the deployment of synchronous optical networking (SONET) and synchronous digital hierarchy (SDH) ring topology [6].

Similar to mobile radio access technology, mobile backhaul also has a complex history [6, 105–107]. GSM operators, started decades ago, used a TDM-based (E1/T1) backhaul network, and invested in plesiochronous digital hierarchy (PDH) multiplexers for aggregation of voice channels. Even though successful, GSM was not designed to support a high volume of data in mobile networks. With the rapid growth of Internet data traffic, 3G/WCDMA was developed as a data-optimized mobile technology and asynchronous transfer mode (ATM) became the converged transport technology for both voice and data, because it supports both fixed bandwidth and best-effort connectivity services. ATM switches were added at the aggregation sites and legacy TDM was replaced by ATM. Soon, the mobile industry came up with LTE and the data community preferred IP/MPLS to ATM because ATM networks require too much manual provisioning and they do not offer a scalable solution capable of supporting 3.5G (HSPA+) and 4G networks throughput. Hence, IP became the connectivity technology of choice for LTE, and mobile backhaul started its migration into IP-optimized transport infrastructure such as Carrier Ethernet and MPLS to support QoS, scalability, and efficient OAM.

LTE-A introduces new features and services with more stringent synchronization, delay, and jitter requirements. Further, intercell communication is growing and more advanced radio signal processing techniques are coming in which put a lot of load onto the network, especially with centralized architectures. The performance–cost trade-off should be considered too, in addition to the technical requirements. These are all challenging the current mobile backhaul and motivating further research in this area. Specifically, it is important to ensure that the mobile backhaul network is not a barrier to the evolution of LTE-A.

Scalability, flexibility, and simplicity are three principal requirements for mobile backhaul solutions. Scalability allows the backhaul network scale out smoothly to support increasing numbers of cell sites at higher capacities. Flexibility and diversity in backhaul access types are important to leverage the closest backhaul access that can meet QoS and cost requirements, particularly in a HetNet and multi-RAT environment. Additionally, to minimize the total cost of ownership, simplified operations are required to allow efficient network deployment and maintenance.

Based on their throughput and latency, the 3GPP has divided backhaul access technologies into two groups: ideal and non-ideal [108]. Non-ideal backhaul refers to typical backhaul access, such as xDSL and microwave, which is widely used in the market. An ideal backhaul, such as dedicated point-to-point connection using optical fiber, has very high throughput and very low latency. A categorization of ideal and non-ideal backhaul and corresponding latency and throughput is listed in Table 6.3.

Being successfully included in 2G/3G radio standards, self-organizing networking (SON) techniques are now mature enough to be applied to other network domains, such as backhaul. Applying SON concepts can simplify planning, reduce operational costs, and improve users' experience [109–111].

6.3.2 Fronthaul

Fronthaul is associated with a new type of RAN architecture consisting of centralized BBUs and one or multiple distributed RRHs installed at remote cell sites located

Table 6.3 Ideal and non-ideal backhaul access technologies

Backhaul type	Access technology	Latency (one way)	Throughput	Priority
Non-ideal	Fiber 1	10–30 ms	10 M–10 Gbps	1
	Fiber 2	5–10 ms	100–1000 Mbps	2
	Fiber 3	2–5 ms	50 M–10 Gbps	1
	DSL	15–60 ms	10–100 Mbps	1
	Cable	25–35 ms	10–100 Mbps	2
	Wireless	5–35 ms	10–100 Mbps	1
Ideal	Fiber 4	Less than 2.5 µs	Up to 10 Gbps	1

kilometers away. This new configuration creates a new transport segment between the distributed RRHs and the centralized BBUs which is called the fronthaul.[4] Traffic is then backhauled from the BBUs to the IP core or evolved packet core. Different from the backhaul network, which can be Carrier Ethernet or various flavors of MPLS for example, the fronthaul network is an optical transmission network. This is because the digitized RF signals aggregated from various RRHs can create huge amount of data (up to 10 Gbit/s), which demands fiber as the transmission medium.

More precisely, the fronthaul transmission is based on digital radio over fiber (D-RoF) technologies such as common public radio interface (CPRI) or open base station architecture initiative (OBSAI) [102, 112, 113]. CPRI is a digital interface standard for high-bandwidth serial data links, derived from cooperation among various radio vendors including Ericsson, Huawei, Alcatel Lucent, and NEC, to name a few. It defines high-bandwidth (up to 10 Gb/s) [114], low latency, and reliable (near zero jitter and bit error rate) transport. This transport system is agnostic to the radio technology because it transports an analog signal in a digital form; therefore, it can be used for 2G, 3G, or 4G and even WiMAX. A number of transport options including dedicated fiber, optical transmission network (OTN), passive optical network (PON), microwave, and wavelength-based systems are potential candidates for CPRI transport.

Fronthaul distance (the maximum distance between RRH and BBU) is limited due to the latency requirement and the timing requirement of hybrid automatic repeat request (HARQ), the protocol which is used as a retransmission mechanism between UE and eNB in LTE networks. Further, to make CPRI a reality, baseband compression will most likely be required and is a current research topic. For example, CPRI has not sufficient capacity to support 20 MHz LTE for a three-sector site with four antennas per sector. LTE-A will require substantial increases in CPRI capacity or signal compression, to support eight antennas per sector and up to five times bandwidth [114].

---

[4]Fronthaul is a new term in the mobile communications' jargon. It is the same as the backhaul access in the old terminology.

6.4   Trends in Wireless Network Densification

Since the invention of the radio up to the present time, wireless system capacity enhancement can be attributed, in decreasing order of impact, to three main factors: increasing the number of wireless access nodes, using additional radio spectrum and improving system spectral efficiency. The above factors contributing to the growth of wireless capacity can be viewed under the common umbrella term of *network densification* [115].[5]

To better understand how these key factors affect the performance of a cellular system, we take a look at the fundamental capacity limit of the simple single-user additive white Gaussian noise (AWGN) channel. This is basically the well-known Shannon capacity upper bound which is given by [117]

$$C = W \log_2 \left(1 + \frac{S}{N}\right), \tag{6.1}$$

where C is the capacity (bits/s), W is the bandwidth (Hz) available for communication, and S and $N = N_0 W$ are the signal and noise powers, at the receiver. This equation assumes one transmitter and one receiver, though multiple antennas can be used in diversity scheme on the receiving side. The throughput of a user in a cellular system, where n users share the bandwidth W and thus interference comes in, is upper bounded by [115]

$$R < C \approx m\frac{W}{n} \log_2 \left(1 + \frac{S}{N + I}\right), \tag{6.2}$$

in which the integer number m (spatial multiplexing gain) denotes the number of spatial streams between the transmitter and user device(s), and S, I, and N denote the signal, interference, and noise power, respectively.

In the high SNR ($SNR \gg 0$) capacity $C \approx W \log_2 \dfrac{S}{N_0 W}$ is logarithmic in signal power and approximately linear in bandwidth. Therefore, network capacity linearly scales with W and increasing the available spectrum is an easy approach for network densification. In practice, however, the spectrum allocated for commercial cellular systems is limited. Then, increasing the spatial multiplexing gain (m), reducing the base station load factor (n), and interference reduction are other factors that can help increase the network capacity; these all can be seen as different spatial densification methods [115], but we study them in different sections of this chapter.

---

[5]In the narrow sense, network densification refers to increasing the number of antennas (access points) per unit area [116]. However, in the broad sense, it is referred to any wireless capacity-enhancing technics [115]. We use the latter definition in this book.

6.4.1   Increasing Spectral Efficiency

In view of limited spectrum available for commercial cellular systems, during past decades, there has been enormous research focus on increasing *spectral efficiency* of wireless systems in order to keep up with exploding capacity needs of customers.

Multiple-Input Multiple-Output (MIMO) Technology

MIMO techniques are the most important advance in recent wireless systems. This technology is now critical part of important standards such as LTE (and LTE-Advanced), WiMax, and different flavors of IEEE 802.11. We discuss traditional point-to-point (single-user) MIMO, as well as two recently developed MIMO systems here.

Point-to-point MIMO: As a key in the design of high-capacity cellular communication systems, MIMO is used for multiple purposes, including to increase throughput (with multiple streams), to increase link range and lower interference (with beamforming), and to improve data integrity (with coding, preconditioning, diversity). In an ideal MIMO system, the data throughput increases *linearly* with the minimum number of transmit and receive antennas [117], which can lead to significant performance gains. However, this limits MIMO benefits for low-complexity cell phones with a single or small number of reception antennas.

Multi-user MIMO: To overcome the limit of single-user MIMO, multi-user MIMO (MU-MIMO) lets multiple users, each with one or more antennas, communicate with each other. Put simply, in an MU-MIMO system, a base station communicates with multiple users. As a result, even if each UE has a single antenna, the sum capacity of MU-MIMO on the downlink can scale linearly by the number of users [118]. Hence, MU-MIMO can achieve MIMO capacity gains just with a multiple antenna base station and a bunch of *single-antenna* UEs. The latter is of particular interest since having multiple antennas is limited on handheld devices. What is more, MU-MIMO does not require a rich scattering environment and, compared to the point-to-point MIMO, it needs a simpler resource allocation as every active terminal utilizes all of the time–frequency bins [119, 120]. Its performance, however, relies on precoding capabilities, and without precoding there is no advantage in MU-MIMO. This technology has been discussed extensively in 3GPP LTE-A.

Massive MIMO: Massive MIMO [121] is a technology that uses a very large number of antennas (e.g., hundreds) that are operated coherently and adaptively. This is a clean break with traditional MIMO systems which may have two, four, or in some cases even eight antennas. Massive MIMO has a large potential to increase the capacity of wireless networks. This increase results from the aggressive spatial multiplexing used in massive MIMO [119]. In fact, abundant antennas help focus energy into even smaller regions of space to bring unprecedented improvements in throughput and energy efficiency. Other benefits of massive MIMO include its potential to significantly reduce latency on the air interface and to simplify the multiple access layer.

Non-Orthogonal Multiple Access (NOMA)

Non-orthogonal multiple access (NOMA) is a technique that allows multiple users to share the same time and frequency, unlike traditional multiple access techniques such as frequency division multiple access (FDMA) and time division multiple access (TDMA) in which users have their own resources. NOMA can be realized in the power domain [122, 123] or in the code domain [124]. Code domain NOMA uses spreading sequences for sharing the resources. On the other hand, power domain NOMA exploits the channel gain differences between the users for multiplexing via power allocation. NOMA is mainly seen as a means to scale up the number of users served in 5G networks both in the uplink [124] and downlink [125]. As such, it can be tailored to typical Internet of Things (IoT) applications in which a massive number of devices sporadically try to transmit small packets. However, NOMA can also increase the spectral efficiency and reduce latency [126, 127]. It is worth noting that the basic theory of NOMA in power domain has been around for many years. The new wave of research on NOMA is motivated by the advance of processors which make it practically implementable.[6]

6.4.2   Interference Mitigation

The performance of wireless networks is inherently limited by their own interference. As such, interference has been the subject of a lot of research focuses on wireless communication, since decades ago. Unfortunately, despite years of intensive research, optimal uplink and downlink transmit/receive strategies for multi-cell networks are unknown, even for the simple case of the two-user network. It is known that NOMA-based techniques result in a superior rate region in multi-cell networks when compared with orthogonal time/frequency allocation [127, 128]. NOMA-based techniques are optimal for single-cell networks, both in the uplink and downlink [127]. Nonetheless, orthogonal resource allocation has been used in the cellular networks from 1G to 4G to avoid intracell interference which makes the transmitters and receivers simpler. Besides, since optimal strategies for multi-cell networks are not well-understood, intercell interference is simply treated as noise.

Interference can be, however, canceled or mitigated by changing antenna patterns in a desired manner. Such systems, known as *adaptive array antennas* or *smart antennas*, are used to locate the main lobe on a specific user or to create a null in the direction of an interferer; they are now replaced by MIMO systems. Other interference reduction techniques include interference alignment, multi-cell processing, or multi-user MIMO.

Although these techniques can considerably improve the spectral efficiency, to cope with the exponentially growing wireless data traffic, network densification in

---

[6]Based on Moore's law processing power increases about 100 times every 10 years.

terms of increasing the number of antennas per unit area seems inevitable. Cell-size shrinking, or cell splitting, is a long-standing approach for this purpose, and the deployment of small cells and HetNets in general, as discussed in Sect. 6.5, is a new manifest of that. Cell shrinking, however, comes at the cost of additional equipment and increased interference. A seemingly simpler alternative is to use very large antennas arrays at the BS [119].

6.4.3 Millimeter Wave Cellular Systems

Millimeter wave (mmW) technology (frequencies between 10–300 GHz[7]) is another frontier for future cellular systems. This is an alternative for more spectrum and offers more than 200 times the spectrum than current cellular allocations [129]. Besides, due to the very small wavelengths of mmW signals and advances in low-power RF circuits it is possible to place large numbers (≥ 32) in small dimensions, and thus benefit from MIMO gains.

The feasibility of mmW cellular communication, however, requires careful assessment, as there are several technical challenges. It is required to deal with the channel impairments and propagation characteristics of the high frequency bands, where free-space path loss is much higher and scattering is less significant. While the former reduces the cell range and SNR, the latter decreases the available diversity, since non-line-of-sight paths are weaker, and results in more blockage and coverage holes. Shadowing can be much more severe in mmW signals, and device power consumption to support large numbers of antennas is high and challenging [129].

6.5 Small Cells and HetNets

Despite implementation of advanced communication techniques such as MIMO in LTE technology and increasing use of Wi-Fi offload to increase network capacity, operators cannot keep up with exploding capacity needs of customers. Therefore, the use of cells with smaller radius serving a smaller number of users appears to be necessary to scale network capacity, particularly in dense urban areas where capacity expansion is most severe.

Small cell is an umbrella term for low-powered, short-range radio access nodes with a typical coverage range of 10 to several hundred meters. Such cells are "small" in comparison to high-power *macrocell* whose typical coverage range can be up to tens of kilometers. They were originally envisioned to provide better voice coverage inside the buildings, where either signal quality was poor or capacity was insufficient,

---

[7]The mmW spectrum is defined as the band between 30–300 GHz, but in the industry it has been loosely referred to any frequency above 10 GHz.

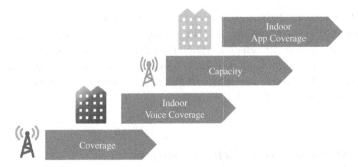

Fig. 6.5 The evolutionary role of small cells in cellular networks. The next and/or current wave of operators' opportunities is shown in *green*

while they are now mainly used as data traffic offloading means from the macrocells [24]. The coverage-filling heritage of small cells is somewhat outdated, though. Operators who still only look at this technology from that point of view are restricting their service opportunities and revenues. Today, small cells are considered more as a business-led solution for launching new services rather than an engineering-led solution for covering network holes. Figure 6.5 illustrates the evolutionary role of small cells in cellular networks.

It is worth mentioning that originally the term *femtocell* was used to describe indoor coverage, and it included *picocell*, being used for enterprise/business premises, and *metrocell*, being used for public/outdoor spaces. Later with the expansion of the scope femtocell technology, the term small cell was adopted to cover all aspects [24]. Today, the term small cell is used to cover *femtocells*, *picocells*, *metrocells*, and *microcells*, in an increasing order in size, i.e., from the smallest to the largest. With such a complex history, there is not a consensus about the naming convention and use cases of different types of small cells. Here we use the definitions of the small cell forum in [21]:

- **Femtocells**: Primarily deployed in consumer and enterprise environments (home and small business). Femtocells are user-deployed access points, and they use consumers' broad connections such as digital subscriber line (DSL), cable, or fiber for connection.
- **Picocells**: Typically deployed in indoor public areas such as airports, train stations, shopping malls, stock exchanges, etc., picocells are operator-installed cells.
- **Metrocells**: Compact BSs, that are unobtrusively deployed in urban areas, usually for capacity alleviation. Metrocells can be mounted on lampposts, located on the sides of buildings or found in stadiums and other hotspots [130].
- **Microcells**: Usually deployed in urban areas or where the footprint of a macrocell is not necessary. The range of a microcell is typically less than two kilometers.

In addition to the above types of small cells, relay nodes and distributed RRHs of a macrocell, which are mounted outside the conventional macro BSs, are also considered as small cells [131].

The abovementioned low-power nodes overlaid on a typically high-power macro base station are deployed to eliminate coverage holes in outdoor and indoor environments, increase the spectral efficiency and network capacity. Such a network that consists of a mix of cell types and radio technologies working together is called a *heterogeneous network* (HetNet). In addition to offering scaling capacity with the number of nodes, HetNets bring access point closer to the user and thus they can efficiently support low mobility and high rate traffic, and lower transmission power to saving energy at the access point and battery life at clients. HetNets also enable seamless integration of unlicensed LAN/PAN technologies into cellular networks. There are a number of technical and deployment challenges facing small-cell networks. Some important ones are listed below:

Backhaul: Diverse deployment scenarios of small cells bring unique challenges for backhaul, making it one of the hottest topics in the small-cell community. Providing the right backhaul solution for each case is critical, especially in public access small cell. Further constraints are caused by operators need for TCO reduction, which usually forces them to leverage existing backhaul capabilities.

Self-organization: Sice some small cells, such as femtocells, are user-deployed without operator supervision,[8] their proper operation depends on their self-organizing features. For example, they need to constantly monitor the network status and optimize their settings to reduce interference.

Handover: Handovers are essential for a seamless service when users move in or out of the cell coverage, and traffic load balancing. It, however, comes at the expense of system overhead, which can be significant in small-cell networks due to the large number of cells and the different backhaul solutions for each cell type. In small-cell networks, unlike single-tier cellular networks, the cross- and intra-tier interference problems are very challenging.

Interference: The deployment of small cells overlying the macrocells creates new cell boundaries, and thus increases the probability that users suffer from intercell interference, which in turn degrades the performance of the overall network. This becomes more critical in hyper-dense small-cell deployments.

Nevertheless, expected to account for a steadily increasing proportion of the offloaded traffic, small cells have already become the most commonly used node for cellular access. By March 2014, 8.4 million small cells, including 8.1 million residential femtocells, were shipped, and at least 64 operators commercially used small cells in their networks [132]. By February 2016, 14 million small cells were shipped. Small cell revenue is projected to be $6 billion in 2020 [133].

Today, there are multiple architectures for small-cell networks, depending on the deployment scenario. In future, small cells will see new architectures where some

---

[8]Note that a centralized radio resource allocation of multi-tier networks would ensure a perfect coexistence between the macrocell and the femtocell networks, but it is highly complex. Hence, a distributed resource allocation is more desirable.

functions in the protocol stack, e.g., packet data convergence protocol layer, that are typically co-located with the RF subsystem of the BS will move to a central location, such as the macrocell [134].

6.5.1 Public Wi-Fi

It is estimated that 90% of the mobile traffic is generated indoor and at hot spots [135]. There are several different indoor radio coverage techniques. Indoor traffic can be handled by macro/microsites, pico/femtocells, repeaters, DASs, and radiating cables [135]. In addition to radio coverage for indoor, Wi-Fi offload is another solution for indoor traffic.

Although mobile network operator may consider Wi-Fi as a competitor for small cells, Wi-Fi and small cells are expected to evolve in parallel rather than compete with each other to offload a steadily increasing proportion of the traffic. Today, most of the smartphones have built-in Wi-Fi receivers, which can automatically switch to Wi-Fi where it is available. The view that Wi-Fi is the most directly competing solutions for the mobile network operator is somewhat outdated. Some mobile operators encourage their smartphone customers to use Wi-Fi where possible. For example, AT&T has provided more than 20,000 free access Wi-Fi in public hotspots, to offload part of data traffic from their mobile network, and improve network performance for those who need it most.

Wi-Fi offload often can provide a faster data service than the outdoor mobile network. However, compared to LTE small cell, Wi-Fi is most suitable for application with a smaller number of users, such as the home or small enterprise. The scheduling and network management becomes a bottleneck for Wi-Fi in high user density scenarios [136]. In contrast, the LTE air interface is well-designed to support a large number of users with guaranteed security and QoS. Therefore, LTE small cell is more suitable for the application in the cellular network deployed by an operator or a large-scale enterprise [136]. Today Wi-Fi is always selected when available. This is not, however, optimal from resource utilization and average user experience. Self-organizing networks can do load distribution between the 3GPP and Wi-Fi for best use of resources and better average service.

6.6 Distributed Antenna Systems (DAS)

A distributed antenna system (DAS) is a network of spatially separated antennas connected to a common source via a transport medium. The antennas are relatively small and serve as repeaters to provide wireless service within a specific geographic area or structure. DAS systems are essentially transport systems that take RF signals from one location and transport it to another. Distributed antenna systems may be deployed indoors or outdoors. An in-building DAS is another way of providing coverage inside large buildings. The idea is to split the transmitted power between separated antenna located in isolated spots of poor coverage, e.g., on different floors, to provide homogeneous coverage.

Distributed antenna systems can be passive or active [135]. In a passive DAS, different components (such as coaxial cables, splitters, taps, attenuators, etc.) are used to split the signal power between the antennas. It basically runs the output power (signal) of a BS through cables to many separate antennas throughout the building. Although passive DASs successfully used in GSM, in higher frequencies the signal degradation can greatly affect the quality and becomes the main issues. Another issue with passive system appears in huge buildings, installation of coaxial cables is not feasible, due to high signal loss along long distances. Also, coaxial cables are hefty and rigid so their installation can be difficult and expensive. In an active DAS, the signal is passed through fiber cables, and long the way, the systems can amplify signals as needed. An active DAS uses different active elements such as the master unit and the remote unit [135]. It contains amplifiers and converters to control all the signals delivery and adjust the signal levels as required.

6.6.1   DAS or Small Cells?

When small cells started appearing, DAS systems have been extensively deployed in large venues. Small cells were thought to make DAS obsolete since they are inexpensive and easy to deploy, but this did not happen and is not likely to happen because these two technologies have some fundamental differences, despite their similar functions, that ensure continuity for both of them.

DAS systems are basically used to transport RF signals from one location to another via fiber. That is, RF signals are modulated onto optical signals and transported to different places before being converted to RF for transmission. Hence, multiple operators can share DAS systems to reduce the cost of deployment and operation. Besides, the DAS remote head-ends where the optical signal is converted into RF are multiband, multicarrier modules. Further, DAS is oblivious to the air interface technology; they can handle 2G/3G/LTE and even Wi-Fi, concurrently. On the other hand, small cells are a single-operator play and when covering inside buildings they have much less RF power than DAS. Sharing small cells is not as easy as DAS and operators resist more.

To summarize, although DAS have been overshadowed by the small-cell approach to indoor coverage, the technique remains relevant in many situations, and operators adopt a mixture of approaches to signal reception issues. These two technologies are complementary to each other to improve coverage and capacity.

6.7   Network Sharing

In the mobile industry, the term network sharing refers to technology that enables two or more mobile operators to use a single infrastructure together. Over the course of the past decade, network sharing has evolved from a mildly interesting concept to a mission critical issue for the mobile industry [137]. Network sharing is stimulated by the same moral as cloud RAN. That is, revenue generated by the data traffic explosion is not increasing the same way. In addition, mobile operators face the challenge of rolling out nationwide LTE networks, using a new technology and new network architecture. They can make substantial savings in capital and operational costs through network sharing.

Network sharing is not a new concept in the mobile communications. Operators throughout the world, even in 2G networks, used to share transmission towers, sites, accommodation, power and air conditioning, etc. In France, for example, Orange shares 40% of sites with other operators in rural areas [138]. However, most of the network sharing agreements in 2G/3G networks were limited to *passive sharing*, in which operators share the basic civil engineering resources. On the contrary, *active network sharing* which is about sharing network equipment such as BS, antennas, radio network controller [139], and potentially radio resources have not been widely deployed in 2G and 3G.

Passive network sharing is known as a means to avoid network duplication, reduce upfront investment costs, and minimize the impact on the environment. As such, regulators in most countries embrace passive network sharing. Active network sharing, on the other hand, is a contested issue and involves technical, commercial, and regulatory complexities. However, as discussed in Sect. 2.5, *deeper sharing* of this sort brings additional benefits and the need to increase savings is creating the pressure to overcome these barriers [137]. Meanwhile, attitudes toward sharing are obviously changing. Advantages and challenges of network sharing solutions are detailed in [139].

6.7.1   Challenges for eUTRAN Sharing

In evolved UMTS terrestrial radio access (eUTRAN) sharing, operators share the active electronic network elements such as BS and the shared eUTRAN is connected to each operator core network. A successful network sharing deployment should, however, address a variety of challenges:

- **Technical issues** such as how to separate traffic between operators, and quality and service differentiation level and methods.
- **Regulatory issues** including negotiation with the regulatory body to adapt license conditions such as spectrum sharing.

- **Commercial and legal aspects** like service-level agreement (SLA), penalties, scope and duration, expenditure split, etc. needs to be defined and agreed on.
- **Process issues** such as management of the shared network.

6.7.2 Standards Specifications

Early standards such as GSM lacked significant support for network sharing. With increasing interest in network sharing solutions, the 3GPP Release 6 introduced a set of basic requirements, which were satisfied by new features in Release 6 for UMTS, Release 8 for LTE, and Release 10 for GSM/EDGE [137]. As network sharing becomes widespread, the 3GPP is developing a number of new capabilities to provide operators with more flexibility and control on the shared networks, and to enable a tighter technical and commercial arrangement between them.

As of Release 11, the 3GPP specifications include two documents dedicated to network sharing matters, in addition to a number of supporting features in other specifications. These are summarized in Table 6.4. Specifically, TS 23.251 elevates

Table 6.4 The 3GPP network sharing specifications

Standard	Specification description
TR 22.951	Service aspects and requirements for network sharing
TS 23.251	Architecture and functions needed to allow multiple core network operators to share a single radio access

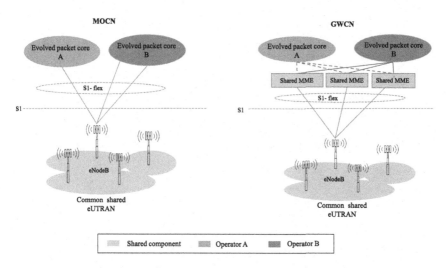

Fig. 6.6 The 3GPP approaches for eUTRAN sharing

two approaches to sharing an LTE eUTRAN, with differences in the core network aspects. These are called the multi-operator core network (MOCN) and gateway core network (GWCN) approaches. In the former, core networks are separated and each network operator has its own EPC. Its benefits include service differentiation, interworking with legacy networks, and falling back to circuit-switched voice services. As depicted in Fig. 6.6, in the GWCN approach, the operators also share the mobility management entity (MME) of the core network. This approach enables additional cost savings compared with the MOCN, but it is less flexible, and reduces the level of differentiation between operators.

6.8   Summary

The access network of cellular networks has evolved largely and adopted numerous new technologies from 1G to 4G, in response to new demands and business and models. This has profoundly affected the standards and network topology. In this chapter, we reviewed the evolution of RAN during the past years and identified new trends in network densification. This includes the evolution of the BSs, BS controllers, and backhaul network. In particular, we discussed why and how the old-stile one-rack BS has moved to the distributed BS with separate BBU and RRH. This sets the stage for new RAN in 5G (cloud RAN) in which BBU and RRH can be spaced miles apart, and will be discussed in the next chapter.

Chapter 7
Cloud RAN

7.1 Cloud RAN: Definition and Concept

Cloud RAN (C-RAN[1]) offers a revolutionary approach to deployment, management, and performance improvement of cellular networks, and is considered as one of the enablers of 5G mobile networks. In principle, C-RAN is a RAN architecture that changes the traditional RAN architecture in a way that it can take advantage of technologies like *cloud computing*, *virtualization*, *multi-user MIMO*, and *software-defined networking*. In a C-RAN architecture conventional distributed cell site base stations are replaced with one or multiple clusters of centralized virtual base stations and each can support up to hundreds of RRHs, placed in different geographical locations in order to provide full coverage of an area. This is achieved by *centralizing RAN* functionality into a shared resource pool or *cloud*. By consolidating individual base station processing into a *single* or *regional* server room, C-RAN offers a compelling value proposition to operators. It provides a clear path to CAPEX/OPEX reduction, easy resource scaling, higher resource utilization, and less energy consumption. Figure 7.1 illustrates the vision of the C-RAN.

The evolution of distributed base station architecture comprising BBU and RRH connected by a fiber optical cable heralds the advent of cloud RAN. Inheriting could technologies benefits, discussed in Chap. 2, to reduce total cost of ownership is the main driver of C-RAN. This is primarily achieved by effective use of physical resources. To this end, BBU centralization is the first but the most important step. BBU virtualization is the next, and probably the most challenging, step which allows decoupling the hardware and software and enables using *commercial servers* and exploiting *statistical multiplexing*. Before discussing the benefits of C-RAN, we take a closer look at these methodologies (concepts).

---

[1]Note that, depending on the source, C-RAN may be interpreted as "Cloud-RAN," "Centralized-RAN," "Consolidated-RAN," "Cooperative-RAN," or even "Clean-RAN." Figure 7.1 illustrates why C-RAN is clean, cooperative, and centralized as well.

© Springer International Publishing AG 2017   87
M. Vaezi and Y. Zhang, *Cloud Mobile Networks*, Wireless Networks,
DOI 10.1007/978-3-319-54496-0_7

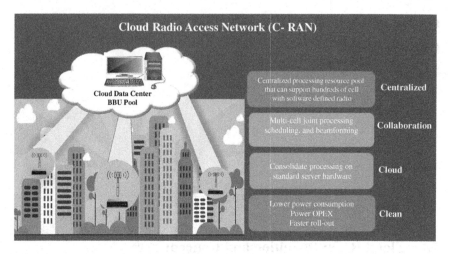

Fig. 7.1   Future cloud radio access networks (C-RAN)

7.1.1   Centralization

Site acquisition for base stations is an expensive and lengthy process, particularly in urban areas. Acquiring real estate together with the civil works comprise a major part of the cost of a new base station deployment. As mentioned in Chap. 1, site acquisition, civil works, construction, and installation, on average, account for about 40% of the total CAPEX of a mobile network, and more than half of the CAPEX in RAN. Centralization of the base stations significantly reduces these expenditures as well as the cost of supplementary equipment, such as power and air-conditioning equipment, required in each site [140]. Leaving the cost of site rental and equipment aside, by consolidating many base stations in one location, centralized RAN brings down the operational and maintenance cost to a great extent. More importantly, centralization opens the door to effective BBU resource sharing via virtualization.

7.1.2   Virtualization

As a disruptive architecture, cloud RAN is much more than just centralizing BBUs in one room and connecting RRHs located at the cell sites. It is an attempt at decoupling the hardware and software platforms of radio base stations [141]. That is, instead of having dedicated hardware for each base station, commercial servers would be deployed in data centers to run base station functions (Fig. 7.2). Recall from Chap. 2 that *virtualization* decouples the software from hardware resources in a way that multiple applications can be run on the same hardware. Hence, due to centralization, commercial servers can be deployed in data centers to run base station functions, once they are virtualized. This leverages the cost structure of *cloud* in running cellular networks and brings cloud service models to cellular networks. Effective use of physical BBU resources due to statistical multiplexing, as explained in the following, is one of the main advantages of base station virtualization.

7.1.3   Statistical Multiplexing

In networking, *statistical multiplexing* refers to mixing the sources with statistically varying rates or feeding them into a common server or buffer. In data networks, sources are typically *bursty* [142, 143], i.e., there are periods when they generate bit or packets at a very high rate compared to the average rate. However, it is a very unlikely scenario, due to statistical independence, that all sources will be concurrently on high rate period at the same time. Hence, the server can be designed at a rate corresponding to the maximum sum rate of all the sources, rather than the sum of the maximum rates of the sources. Obviously, this results in a better resource utilization, if the sources have different statistics. Put differently, statistical multiplexing enables sharing a resource on a demand basis, and thus more efficiently than a priori allocating a fixed portion to each user. Statistical multiplexing gain can be defined as

$$\eta = \frac{\sum \max r_i}{\max \sum r_i}, \tag{7.1}$$

where r_i is the instantaneous rate of source i. The gain increases when the number of sources increases and/or when some or all of the sources highly bursty. For example, in [144] it is shown that multiplexing gain linearly increases with network size and traffic intensity, in a WiMAX network where the medium access (MAC) layers of the base stations are pooled.

Traditionally, resource dimensioning/allocation in cellular networks is on per cell basis, and based on the busy hour traffic requirements. In practice, the cell load varies throughout the day and busy hour traffic can be much higher than the off-peak hours traffic. Also, due to the inherent mobility of the mobile networks users, the peak hour differs from one cell to another, depending on the geographical area they cover. For example, cell sites in commercial areas will experience peak traffic during working hours, while those located in residential areas and suburbs are underutilized. The load becomes reverse in the evening, as people go back home.

With this insight, it is expected that if statistical multiplexing is applied in the resource allocation of a collection cell sites, a drastic reduction in computing/processing hardware is possible. Many independent studies support this notion. For instance, in [145], based on real-world data, it is shown that at least 22% saving in compute resources can be achieved, if the signals of all base stations are processed on a shared pool of compute resources, in a central location. Werthmann et al. [146] show that, in a 10 MHz LTE system, aggregation of 57 sectors into a single sector BS

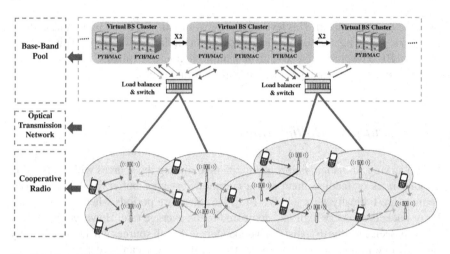

Fig. 7.2 Cloud RAN [8]

can save more than a 25% in compute resources. Other works, such as [147–149], report even much more gain.

7.1.4 Multi-cell Cooperation

A key in the design of high-capacity cellular communication systems, MIMO technology offers significant increases in data throughput and link range for the same bandwidth and transmit power. It is well-known that, under ideal conditions, the data throughput increases *linearly* with the minimum number of transmit and receive antennas [117]. Further, even if the UEs have a single antenna, some capacity of multi-user MIMO (MU-MIMO)[2] on the downlink can scale linearly by the number of users [118]. This means that MU-MIMO can achieve MIMO capacity gains just with a multiple antenna base station and a bunch of *single-antenna* UEs. The latter is of particular interest since having multiple antennas is limited on handheld devices. Additionally, MU-MIMO does not require a rich scattering environment and, compared to the point-to-point MIMO, it needs a simpler resource allocation as every active terminal utilizes all of the time–frequency bins [119]. Its performance, however, relies on precoding capabilities, and without precoding there is no advantage in MU-MIMO.

In an attempt to further increase the data throughput, upcoming cellular standards like the 3GPP LTE-A are aiming at universal frequency reuse. Simultaneous transmissions on the same frequency by neighboring base stations could, however, lead to high levels of *intercell interference*. In dense networks, where interference

---

[2]In an MU-MIMO system, a base station communicates with multiple users.

emerges as the key capacity limiting factor, *multi-cell cooperation* can dramatically improve the system performance [127, 150–152]. Literally, mimicking the benefits of a large virtual MIMO array, multi-cell systems exploit intercell interference by jointly processing the user data through several interfering base stations.

In practice, the performance of multi-cell systems is limited by a variety of non-idealities, such as insufficient channel state information (CSI), high computational complexity, limited level of cooperation between the BSs (e.g., due to finite backhaul capacity), and transceiver impairments. Besides, the performance of multi-cell systems is highly dependent on resource allocation, i.e., how to divide the time, power, frequency, and spatial resources among users [153]. Hence, in a cellular network where multiple base stations coordinate in their resource allocation strategies, but transmit and receive data streams independently, the following are critical and should be addressed jointly [153]: *(1) beamforming*: to allocate suitable transmit/receive beams at the base stations and UEs, *(2) scheduling*: to determine which user should be served in each frequency/time slot for each beam, and *(3) power spectrum allocation*: to identify the appropriate power spectrum for each beam. Such an optimization must be performed separately for the uplink and the downlink and repeatedly over time, as wireless channels vary very fast.

Multi-cell systems known as *coordinated multipoint (CoMP)* transmission and reception in the 3GPP LTE-A. CoMP is a complex set of techniques that enable the dynamic coordination of transmission and reception over a variety of different base stations. It primarily targets intercell interference reduction, particularly at the cell-edges, but can also be used to improve the network coverage and utilization as well as the user quality of experience (QoE). Four scenarios were listed in the 3GPP LTE-A study of CoMP. These consider coordination between [63, 151, 154]: (1) the cells (sectors) controlled by the same macro base station (2) the cells belonging to different macro base stations in the network (3) a macro cell and its distributed antennas—all with the same (macro cell) identity (4) a macro cell and a small cell within its coverage—each with their own cell identity.

The first two scenarios are targeted for homogeneous networks whereas the last two scenarios are targeted for heterogeneous networks. These four scenarios differ in terms of gain and requirements. In the first scenario, for example, no backhaul connection is required [63, 151]. Depending on the required backhaul link between coordinated points and the level of scheduling, CoMP techniques can be broadly categorized as [154]

- **Joint processing (JP)**: JP implies that users' data is available at multiple BSs. This can be implemented in two forms: (1) *joint transmission (JT)* which refers to the case where multiple transmission points (TPs) are transmitting the signal to a single UE device on the same time–frequency bin, in a coherent or noncoherent manner, and (2) *dynamic point selection* where transmission is done from one point (serving cell) at a time. The TP with the best link quality is dynamically selected according to the instantaneous channel condition, at each subframe.
- **Coordinated scheduling/beamforming (CS/CB)**: In *coordinated beamforming (CB)* each base station has a disjoint set of users to serve but it selects its transmit

strategies in coordination with other base stations, to reduce intercell interference (ICI). Similarly, with CS the scheduling decision (i.e., which specific UE is selected for transmission) across different TPs are made with coordination among others corresponding cells to reduce intercell interference (ICI) [151]. Note that in both cases data is only available at one TP (the serving cell) but user decisions are made jointly.[3]

CoMP has the potential to turn the into useful signal, especially at the cell borders. By using combining techniques, despite apparently destructive nature of the interference, CoMP can utilize the interference constructively to reduce the levels of interference at UE terminals. The main benefits of CoMP transmission/reception can be summarized as:

- **Better network utilization**: by providing connections to several base stations at the same time, and via passing data through the least loaded base station, CoMP enables better resource utilization.
- **Interference management**: joint transmission has the potential to turn the interference into a useful signal. Also, by appropriate selection of the beamforming weights, coordinated beamforming steers interference toward the null space of the interfered UE terminal and reduces the interference experienced by that terminal [151, 155].
- **Better user experience**: joint reception from multiple base stations using CoMP techniques helps to increase the overall received power at the UE, which in turn, translates into higher throughput and better user experience.[4]
- **Higher capacity**: cooperative communication schemes, generally, improve system capacity and diversity. Specifically, joint transmission can improve the channel rank which results in higher capacity. Besides, joint processing provides macro diversity protection for shadowing channels [156].

Both uplink and downlink CoMP were considered in the 3GPP study [154]. Obviously, CoMP techniques are different for the uplink and downlink. Uplink CoMP tends to have less standard impact [63] since the eNBs are connected to each other, whereas the UEs are not. As a result, processing at the network side can be more transparent than that at the UE. In contrast, downlink CoMP requires more standardization work as it involves CSI reporting from UE, interference measurement, and reference signal design [151].

---

[3]It is worth noting that the above CoMP techniques can be applied to non-orthogonal multiple access (NOMA) methods. A list of such techniques can be found in [127] and the references therein.

[4]User performance is usually measured by mean square error, bit/symbol error probability, or information rate (bits per channel use). These all improve with signal-to-interference-plus-noise ratio (SINR) [153].

7.2 Potentials and Benefits of Cloud RAN Architecture

7.2.1 Potentials of Cloud RAN

C-RAN will have profound strategic implications on the operator–vendor relationship. Operators can upgrade their networks more agilely; they can even select between vendors. New concepts like RANaaS emerge and implementation of RAN sharing and multi-RAT RAN becomes much easier in light of C-RAN. MVNOs can build their networks from a shared pool of resources, which enables them to dynamically use/release resources as per demand. Apart from these strategic implications, there are several measured benefits to cloud RAN from different perspectives, such as reducing the cost of network operations and maintenance, saving the cost of equipment, improving the network capacity, and reducing energy consumption as well as carbon footprint; we will elaborate them in Sect. 7.2.

The evolution of base station architecture with split BBU and RRH connected by a fiber link heralds the arrival of C-RAN. While the availability of fiber is a prerequisite to a positive business case, C-RAN can be viable even if the fiber is not available. However, its cost may be comparable to that of small cells. As a result, even without fiber assets, C-RAN in a HetNet architecture becomes a viable option for wireless operators [141]. We will discuss this in more detail later in Sect. 7.4.

Finally, C-RAN has a potential for green telecom. In addition to reducing OPEX, to expand the network into rural areas where power availability is poor, green communication is important. Green telecom is becoming more viable as renewable energy technology is becoming available at a reducing cost and also for sociopolitical trends toward environmental responsibility.

7.2.2 Benefits of Cloud RAN

The benefits of cloud RAN mainly center on reducing the cost of network deployment and operation but are not limited to that. Here, we summarize major advantages of a C-RAN architecture to traditional RAN.

- **Lower CAPEX**: A major part of CAPEX reduction is simply because of the elimination of the need for individual site acquisition, civil works, installation, and supplementary equipment for each site. Another big saving in CAPEX is archived in light of statistical multiplexing, i.e., due to shared use of pooled resources, which reduces the required compute resources. Further, cheap general-purpose processor hardware can be used instead of custom-made processor hardware.
- **Lower OPEX**: C-RAN will deliver significant reductions in OPEX due to reduced site rental costs, less power consumption, fewer truck rolls, and ease of operation and maintenance (O&M). A major reason for this is simply the aggregation of the BBUs in one location or a few big rooms. This centralization saves a lot of the

O&M cost when compared with much larger number of sites in traditional RAN. For example, instead of having one A/C per site, one or a few of them are enough per center. This alone reduces power consumption to a great extent. In addition, in a C-RAN architecture the functionality of cell site equipment is simpler, as L1/L2 processing moves to the BBU room. Hence, the size, weight, and power consumption of cell site equipment reduces and they can be mounted on less strong poles and on walls with minimum site support, and require minor maintenance.

- **Higher capacity**: C-RAN pools and shares (via virtualization) the BBU resources in a BBU room. Physically located in the same place, virtual BBUs can easily share the signaling, traffic data and channel state information of active UEs to carry out load balancing and advanced features such as network MIMO to exploit intercell interference by jointly processing the user data through several interfering base stations. Through advanced multi-cell cooperation techniques and load balancing, C-RAN is capable of reducing the interference and improving the spectral efficiency and network capacity, particularly in dense networks where throughput is interference limited. These features can deliver up to 2 times the capacity of LTE R8 [16]. The gain varies on the downlink and uplink and also depends on the access method among other factors. The gain can be marginal or substantial and provides a further reduction in the cost of capacity.

- **Reducing energy consumption and carbon footprint**: C-RAN is seen as a road toward green RAN [8, 157], for its great potential in reducing energy consumption and carbon footprint. Potential energy savings can be drawn from a number of areas: (1) multiple BBUs can share facilities, e.g., air-conditioning (2) resource utilization improves due to pooling and sharing, which translates to less hardware for the same traffic (3) BBU resources can be switched off much more easily than distributed hardware when required, e.g., when traffic demand is low (4) multiple radio access technologies and network efficiency are supported more efficiently (5) the density of RRHs can be increased (virtually) due to the use of cooperative techniques for interference management, resulting in a shorter distance from the RRHs to the UEs and less energy consumption for signal transmission.
It is obvious that such an architecture is more environmental-friendly and reduces CO_2 emissions. Green telecom can be driven by several factors including (i) to reduce OPEX (ii) to expand network into rural areas where power availability is poor (iii) renewable energy technology becoming available at reducing cost (iv) sociopolitical trends toward environmental responsibility [42].

- **Scalability and reliability**: Flexible network capacity expansion is another advantage of the C-RAN. This primarily facilitates the compute resource expansion at the BBU room. Access points are small, low-power RRH which can be deployed in the coverage area more conveniently. When base stations deployed in a geographical area are abstracted as a virtual big base station, RAN infrastructure reliability is naturally enhanced and the road is paved for a software-defined RAN, which makes RAN more programmable and flexible [65, 158]. In the C-RAN architecture, radio protocols are implemented using software-defined radios in the cloud. For example, a complete LTE-based station stack can be implemented in software

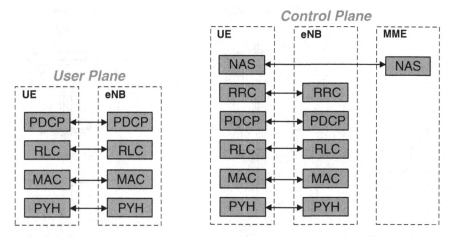

Fig. 7.3 Overall RAN protocol architecture: user and control planes

running on Intel architecture processor, offering the versatility to add new radio protocols simply with a software upgrade.

7.3 Cloud RAN Implementation Scenarios

Before embarking on C-RAN implementation scenarios, with the overall network architecture in mind, we briefly discuss the LTE RAN protocol architecture for the user and control planes, depicted in Fig. 7.3. This is later used to differentiate different C-RAN architects.

7.3.1 RAN Protocol Architecture

Modern telecommunication standards such as LTE, can be divided into three different layers, namely, the physical layer (L1), data link layer (L2), and network layer (L3). Implemented differently in different wireless protocols, each layer may include some sub-layers. According to 3GPP, LTE layer 2 structure consists of the media access controller (MAC), radio link control (RLC), and packet data convergence protocol (PDCP). The basic protocol structure of LTE is illustrated in Fig. 7.4. The reader is referred to [9] for further details.

The access network in LTE is simply a network of base stations, evolved NodeBs (eNBs). To speed up the connection setup and reduce the time required for a handover, the intelligence is distributed amongst the eNBs, and there is no centralized intelligent controller in LTE. From an end user perspective, the connection set up

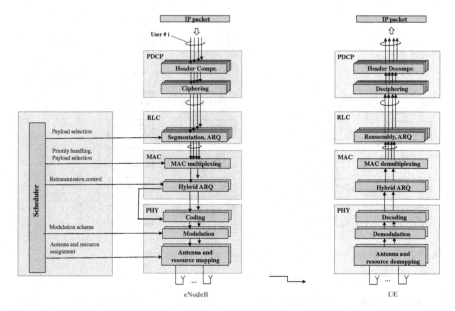

Fig. 7.4 LTE protocol stack (downlink) [9]

time is a crucial QoE criterion [159, 160] for many real-time data session, especially in on-line gaming. Likewise, handover time is essential for real-time services as the end users tend to end calls if the handover takes too long. In the distributed solution, MAC protocol layer[5] is represented only in the UE and eNB, leading to faster communication and decisions between the UE and eNB, whereas in UMTS, for example, the MAC protocol is located in the controller rather than in the NB [161].

The PDCP layer receives packets from the network layer and performs IP header compression, ciphering, and integrity protection of the transmitted data for the control plane. Next to the PDCP, the RLC layer is responsible for error correction through automatic repeat request (ARQ), concatenation, segmentation, duplicate detection, and in-sequence delivery to higher layers. It provides services to the PDCP in the form of radio bearers [9, 162]. The MAC layer handles the mapping between logical and transport channels, multiplexing RLC links, uplink and downlink scheduling and priority handling, HARQ, and feedback creation and processing. MAC layer offers services to the RLC in the form of logical channels.

---

[5]MAC layer is responsible for scheduling which is a key function for fast adjustment and efficient utilization of radio resources. Note that, the transmission time interval (TTI) is only 1 ms, and in this period, the eNB scheduler shall: (1) decide which modulation and coding scheme to use. (The decision is based on the reported perceived radio quality per UE and relies on rapid adaptation to channel variations, employing HARQ with soft-combining and rate adaptation.) (2) prioritize the QoS service requirements amongst the UEs. (3) inform the UEs of allocated radio resources. The eNB schedules the UEs both on the downlink and the uplink.

The multiple access scheme for the LTE physical layer (PHY) is based on orthogonal frequency- division multiple access (OFDMA) and single carrier frequency division multiple access (SC-FDMA) in the downlink and uplink, respectively. PHY layer is responsible for baseband signal representation. The baseband signal processing includes channel coding, rate matching, interleaving/scrambling, modulation, layer mapping, multi-antenna precoding, and generation of OFDM signal for each antenna port [9, 163]. The physical layer provides services to the MAC layer in the form of transport channels.

7.3.2 Proposed Architectures for C-RAN

Centralized RAN appears to be a natural evolution of the distributed BSs,[6] where the BBU and RRH can be spaced miles apart. However, moving from the centralized to cloud RAN may happen in several phases. This is because there are several different ways for splitting up RAN functions between the RRH and BBU. Depending on the function splitting between BBU and RRH, *fully centralized* and *partially centralized* C-RAN are plausible.

In a fully centralized solution, the layer 1 (baseband), layer 2, and layer 3 functions of RAN are located in BBU, whereas in a partially centralized C-RAN, RRH integrates both radio and baseband functions; it may also include some of the higher layer functions.[7] Other higher layer functions are still located in BBU. As a result, different partially centralized architects are possible. Figure 7.5 illustrates how step by step we can move from current RAN to cloud RAN. The first architect shows the current RAN with classical eNB, where L1 and L2 functions are located in cell site and thus are distributed in the network. In the second architect, most of the L2 functions are moved to a centralized BBU, and thus eNB becomes slim. Likewise, in the third architect, in addition to L2 functions part of the PHY functions are centralized which leaves only RF function in the cell site. That is, eNB includes the RRH and some PHY functions of BBU. This is called an extended RRH. Finally, in the fourth architect, eNB becomes an RRH and BBU functions are fully centralized.

As can be seen in this figure, centralizing more functions simplifies the RRH but requires a higher fronthaul bandwidth. The latter is because the split point is getting

---

[6] A distributed or split BS architecture typically consists of a BBU and RRH connected by cabling. RRH which contains the base station's RF circuitry (filter and analog-to-digital/digital-to-analog converters and up/down converter), low noise amplifier, and power amplifier is installed next to the antenna on a tower or rooftop rather than at the tower bottom or a building basement where old-style macro base stations are typically located. By installing RRHs close to antennas, short coaxial jumpers are used for connecting RRH to the antennas; however, operators are required to extend data transmission, power, and grounding cables from the base station to the top of the tower or rooftop [164, 165].

[7] Note that in the later case the BBU does not include some of the baseband functions but it is still called BBU for the simplicity. Similarly, the cell site includes some of the functions of the BBU in addition to the "radio" related functions, but still is called RRH to avoid a different naming. Partially centralized C-RAN can also have different architects.

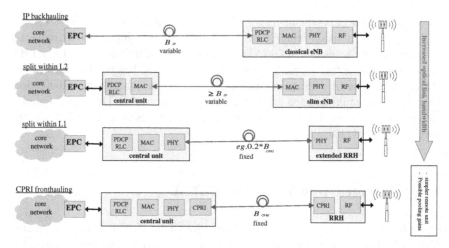

Fig. 7.5 Different C-RAN implementation scenarios [source Alcatel-Lucent]

closer to RF frontend where the data rate is higher due to the aggregation of more layer headers and other baseband signal processing loads, such as channel coding. Fronthaul technology is one of the main challenges of cloud RAN, and the distance from centralized BBU and cell sites varies based on the chosen technology.

RAN functions may also be divided in a different way. Theoretically, this functional split may happen on any protocol layer or on the interfere between the layers. For example, based on timing requirements, the functions can be divided into four subgroups [166]:

1. LTE L1 - PHY
2. LTE L2 (realtime) - MAC, DL RLC, and DL/UL scheduling
3. LTE L2 (non-realtime) - UL RLC, and PDCP
4. LTE L3 - control plane.

Then, depending on where we place the functional split, the implementation of C-RAN and it benefits and challenges will be different. The split point between the centralized and distributed functions is very important as it determines the virtualization granularity. In general, the lower the functional split in the LTE protocol stack the higher the granularity of virtualization. This can increase virtualization and codification gains. However, it makes fronthaul requirement, in terms of latency and throughput, more stringent [167]. Depending on the split point, defined based on the above subgroups, the following C-RAN architects are possible:

- **Architect 1**: All functions are centralized
- **Architect 2**: Only L1 (physical layer) functions are not centralized
- **Architect 3**: Real-time L2 and L1 functions are not centralized
- **Architect 4**: Only L3 (network layer) functions are centralized.

Architect 1 is basically the fully centralized solution whereas architect 4 is the classical RAN implementation. The two other architects are partially centralized solutions. In the following, the four architects are compared by listing their advantages and disadvantages.

The fully centralized RAN enjoys all the benefits of C-RAN, discussed earlier in this chapter. The main challenge is to meet the tight latency requirements of the PHY layer processing to maximize the available time budget for the MAC layer scheduler. Besides, the strict latency and throughput requirements between the L1 and radio frequency integrated circuit (RFIC) make the fronthaul technology, and thus C-RAN, an expensive solution. To satisfy the stringent one millisecond transmission time interval requirement of the LTE physical layer, operators must choose a very low-latency access technology for the L1 to RFIC interface. To be specific, among the access technologies listed in Table 6.3, only the ideal connection (Fiber 4) can be used for this purpose. In addition, the distance between the L1 and the RFIC should be such that allows the latency requirements to be met.

Compared to Architect 1, Architect 2 has less fronthaul throughput. This is because the overhead originated from physical layer processing (such as channel coding, precoding, layer mapping, etc.) does not exist. The latency budgets of Architect 2 are still critical and similar to Architect 1, hence the corresponding disadvantages apply. On top of that, L1 functions are now located in the cell site, eliminating the gains associated to their centralization. For instance, since detection and decoding (and modulation and encoding) are not centralized, we do not benefit from advanced precoding and high computation diversity as much as the case where those functions are centralized [167].

Architect 3, does not centralized time-critical functions of eNB (scheduler, RLC DL, MAC, L1); i.e., they are left in the cell site (RRH). Consequently, latency requirements are less stringent in this architect, which decreases its dependency to expensive fronthaul technology. Hence, operators may use a number of access technologies like gigabit Ethernet and microwave in addition to fiber [166]. On the flip side, since the granularity level of virtualization in Architect 3 is less than that in Architect 2, benefits of virtualization and multi-cell proceeding will be less, too. In addition, L2 functions are divided into two parts, requiring non-3GPP specific communications between these two entities.

Architect 4 only centralizes the L3 functions and is, from the implementation point of view, the less challenging among the four architects; it eliminates the need for custom backhaul technology capable of meeting timing requirements. Since all L1 and L2 functions are executed near the cell site, this scenario eliminates the need for expensive backhaul technology because the C-RAN is just responsible for the control signaling. Additionally, *load balancing* in RAN and fronthaul is possible due to the L3 centralization. Other benefits of L3 centralization are *flexible QoS management* and increase security per radio access point [167]. On the other hand, the L1 and L2 functions are not virtualized, and multi-cell processing gains are not achieved. Overall, this architect brings some capacity gain but as demand grows, the need for more cell sites will be more than the other architects.

7.3.3 RAN as a Service (RANaaS)

RANaaS enables the offering of RAN to mobile operators as a service running on a data center [168], and implies visualizing RAN and implementing the functionality of the eNodeB radio protocol stack in software running on general-purpose processors (GPP). In [167] RANaaS is referred to a flexible trade-off between the fully centralized and traditional implementation of RAN, where RAN functionality is centralized through an open IT platform based on a cloud infrastructure [156, 169]. With latter definition, RANaaS is basically another name for partially centralized RAN in which the level of centralization varies depending on the actual needs and network characteristics, i.e., the functional split is flexible between the cell site and centralized BBU pool. RANaaS has similar challenges and advantages as C-RAN and in general the cloud computing platforms.

7.4   Cloud RAN or Small Cells?

Cloud RAN and *small cells* have both been advertised as cost-saving solutions to manage high-density traffic problem. While cloud RAN is based on putting the baseband processing together and pooling them, small cells distribute the baseband processing over the cell sites. A natural question is then to know which one is better, i.e., shall we bring the baseband processing together, or spread it out? Both technologies offer a way to save cost and provide a boost in capacity, but they come with their challenges as well.

Intended to reduce cost per bit by increasing spectral efficiency and throughput, small cells are low-power, low-cost base stations which bring the overall cost of baseband processing down when compared with macro cells. Effectively, small cells bring access points closer to the users to improve SINR, which in turn increases the spectral efficiency. They can drop the cost by a factor of 4, compared to a macro LTE network [170]. However, they require a high level of coordination to find intercell interference, they need heterogeneous backhaul solutions, and they many be underutilized due to a high level of spatial and temporal traffic fluctuations.

C-RAN makes LTE-A features such as CoMP and enhanced intercell interference coordination (eICIC) easier to implement. This is because these features need a great amount of coordination and locating the baseband processing in one place reduces the complexity of coordination. CoMP results in an added capacity and higher throughput which can be as much as 80% at the cell edge [170]. Further, major savings come from the centralization and pooling of the BBU's, as we saw earlier in this chapter.

In effect, a low-power RRH in C-RAN is similar to a small cell, except that the baseband is in a central remote location which, in this case, can be a macro cell site [141]. While C-RAN has the advantage of greater coordination with the macro cell layer, it brings about challenges in the transmission network, more specifically, in the *fronthaul*, i.e., the link between the BBUs in the central office and the RRHs in the cell sites. It requires constant high-capacity fronthaul (order of Gbps) as opposed to the relatively low-capacity requirements for small cell backhaul (order of 100 Mbps) [141, 170].

So which architecture brings more saving per bit? and, which one will be preferred by the operators? There is no clear-cut answer to these questions. The C-RAN architecture is more supported by Asian operators such as China Mobile, NTT DoCoMo, SKT, and KT, which have access to cheap fiber, whereas the small cell architecture seems to be popular with many mobile operators because the fronthaul cost outweighs the operational savings of the BBU centralization. Nevertheless, note that we are at the early days of C-RAN, where mostly centralization of BBUs has been implemented. With the advance of base station virtualization, the structure will move from merely centralized RAN towards the cloud RAN. When baseband resource sharing is implemented and general-purpose processing hardware is in place for base stations more gains are expected. Also, as networks evolve to incorporate LTE-A techniques and to leverage the HetNets architecture, C-RAN provides an alternative solution to small cells [141]. Although, CoMP and eICIC will be implemented with small cells, their benefit is smaller.

It is clear that if fiber bandwidth is free, C-RAN is preferred to small cells. In a HetNet configuration, C-RAN can be a viable solution for wireless network operators even without having fiber assets; but, the cost of the two architectures can be very similar. In the end, it seems that operators which do not have fiber assets will favor the small cell architecture; they may, however, consider C-RAN architecture for special situations such as a stadium with very high density. To this end, wireless fronthaul solutions need to develop to meet C-RAN requirements for high capacity and stringent latency.

7.5 Cloud RAN Challenges and Research Frontiers

With all the benefits cloud RAN architecture offers, there a number of important challenges and areas for development. We will consider them in the following subsections.

7.5.1 Challenges

In the years to come, we expect to see more developments related to wireless fronthaul and base station virtualization. The corresponding technical issues need to be addressed before C-RAN becomes a reality.

Fronthaul is a new term used to refer to the transport segment between the RRHs and the centralized BBUs in C-RAN configuration (Fig. 7.6); the segment between the BBU pool to the core network is still called backhaul. Currently, *common public radio interface (CPRI)* links are used to connect the RRH and DAS to the BBU to carry

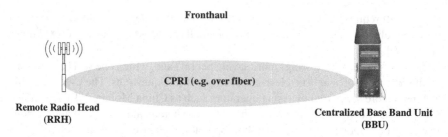

Fig. 7.6 Fronthaul transmission (commonly CPRI is used to connect RRH to BBU)

digitized baseband complex samples (in-phase/quadrature-phase (IQ) components) [114, 171]. CPRI can support high-bandwidth transport up to 10 Gb/s with low latency, and near zero jitter and bit error rate. However, as we saw in Sect. 6.3.2, CPRI requires baseband compression to support high-capacity sites (e.g., macro sites in LTE-A or even LTE networks). Obviously, C-RAN fronthaul is much more demanding in this sense, and thus baseband compression is indispensable if CPRI is used. Also, since the distance between RRHs to the BBU pool varies and usually is very large, jitter and delay become more challenging to tackle, as elaborated below:

- **High bit rates needed at the fronthaul**: The transport network has to constantly carry a large amount of data. Hence, the fronthaul network should have high capacity, e.g., optical links, especially in the fully centralized RAN. This is very costly for the carriers which do not own fiber asset. Even if such links are available to transfer digital complex-baseband wireless signals to/from RRH, technically signal compression is required to make data transfer, from high-capacity LTE/LTE-A sites, over CPRI a reality. In this view, baseband compression is indispensable in C-RAN fronthaul.

- **Latency constrains**: *Latency* is an important factor that affects users QoE. For example, it governs parameters such as the time it takes for a requested Internet page to display or a file to be downloaded. From the user perspective, latency is the time required for a data packet to travel from the UE to the content server on the Internet and back (round trip), as seen in Sect. 6.1.2. It includes delay from propagation, buffering and queuing, transmission, and signal processing that is introduced at the link and network element through which a packet travels [172].

LTE offers a huge improvement in latency values compared with 3G or even HSPA, by reducing the user-plane latency from 30 ms (HSPA) to about 10 ms [172]. Table 7.1 list the data rate and latency requirements for the HSPA+, LTE, and LTE-A [5]. Cloud RAN solutions (including those in fronthaul) must consider stringent real-time requirements of the RAN. For instance, failure to comply with the timing requirement of HARQ[8] induces an unnecessary transmission, thereby

---

[8]HARQ in LTE requires an acknowledgment (positive/negative) 3 ms after receiving a transport block.

Table 7.1 HSPA+, LTE, and LTE-Advanced requirements

Parameter	HSPA+	LTE	LTE-A
Peak downlink speed (Mbit/s)	168	300	3,000
Peak uplink speed (Mbit/s)	22	75	1,500
Idle to connected latency (ms)	<100	<100	<50
Dormant to active latency (ms)	<50	<50	<10
User-plane latency (one-way)	<10	<5	<5

lowering the throughput [140]. In any case, the delay imposed by the compression algorithm should not exceed a certain value, and this cannot exceed the maximum allowable delay of the fronthaul [173]. In practice, using ideal fiber (see Table 6.3) for transport leaves an acceptable delay budget for processing requirements and propagation delay, but it can be very expensive.

- **Jitter and synchronization**: Jitter is the variation in the packet transit latency. It is caused by the variation of the load on the network and is affected by queuing, contention, and serialization on the path through the network. It can also be generated and introduced into the fronthaul network during signal mapping/multiplexing [174]. Jitter can eventually cause errors during the data and clock recovery process at RRH. Since the clock frequency of the remote RRH should be synchronized with the BBU, the RRH must obtain a reference clock by recovering a timing clock from CPRI I/Q bit streams transmitted by BBU,[9] and the accuracy of this clock should be less than ± 2 ppb, two billionth of the reference clock frequency [174]. Jitter becomes more critical in a C-RAN architect as the RRH and BBU can be miles away.

7.5.2 Research Frontiers

CPRI can support high bandwidth transport up to 10 Gbps with low latency, and near zero jitter and bit error rate. But, as Table 7.2 illustrates [175], using LTE-A protocol, fronthaul throughput for a single site can be more than 10 Gbps. This means, even with the current technology, if carrier aggregation is applied, fronthaul data compression is required. Obviously, to make CPRI a reality in C-RAN, baseband compression is necessary. Baseband signal compression is hence a hot research

---

[9]While, the BBU can generate a master reference clock using its GPS receiver, the RRH cannot as it is not equipped with a GPS receiver [174].

Table 7.2 Fronthaul throughput for a 3 sector LTE-A site

Parameter	Setting	Unit
LTE carriers	5	
Bandwidth	100	MHz
MIMO antennas	2×2	Tx-Rx
Bits-per-I/Q	15	Bits
Throughput	13.8	Gbps

topic, and a variety of compression techniques are being currently considered by a number of organizations and operators as well as many different scholars to reduce the fiber bandwidth required for data transmission. It should, however, be noted that the difference between the ideal symbols and the measured symbols after the equalization, which is called the error vector magnitude (EVM[10]), should not exceed a certain level. In LTE-A, EVM requirements for QPSK, 16-QAM and 64-QAM modulation schemes are 17.5, 12.5, and 8%, respectively [176]. We review the baseband compression techniques in the following.

Fronthaul compression techniques can be classified in different ways. In general, *lossless* or *lossy* compression techniques can be used for baseband data compression [177]. Besides, the compression can be in different domains; it can be *temporal*, *spectral*, or *spatial* (see Table 7.3). From an information theoretic perspective, fronthaul compression techniques also can be divided into *point-to-point* or *multiterminal* techniques. The latter case includes joint decompression and/or joint compression. For example, distributed source coding techniques can be used for compression [178]. Looking from another perspective, fronthaul compression techniques can be quantization-based compression or compressive sensing (CS)-based compression and spatial filtering [179].

A low-latency baseband signal compression algorithm, which removes redundancy from LTE signal in the spectral domain is proposed in [171, 180]. The proposed algorithms first remove the redundancy and then perform block scaling, together with a linear or nonlinear quantizer, to minimize quantization error. These algorithms reduce the amount of data transmitted between BBU and RRH; they yield good performance under 1/2 compression rate in a practical propagation environment. Frequency domain compression increases IQ mapping complexity, and thus makes the interface logic design and processing more complex.

More sophisticated fronthaul compression include using distributed source coding (DSC) or joint compression, and are called network-aware compressions. Essentially network-aware compressions provide significant gains with respect to the point-to-point compressions. A DSC-based fronthaul compression, for example, enables the BBU pool to leverage the correlation among the signals received by neighboring RRHs. Therefore, similar to the DSC gain with respect to point-to-point coding

---

[10]EVM is defined as the square root of the ratio of the mean error vector power to the mean reference power expressed in percent [176].

Table 7.3 Fronthaul (backhaul) compression techniques

Compression domain	Techniques	Compression ratio	Error vector magnitude (%)	References
Time	Rescaling, nonuniform quantization, noise-shaping error feedback, resampling	0.33/0.19	< 2%	[114]
	Distributed compression via conditional KLT			[181]
	Lossy/lossless compression	0.25/0.66	NA/0	[177]
Spectrum	Up/down sampling, block scaling, and quantization	0.52/0.39/0.30	≈1.2/1.35/2.4	[171, 180]
Spectral–Temporal	Statistical multiplexing to avoid full utilization of the links in low load situations	Up to 0.03 at 10% load	NA	[182]
Spatial	Based on multiple antennas			

(both in binary [183, 184] and real [185] domains), we can expect a considerable gain in fronthaul compression. The above network-aware compression techniques are, however, available only for uplink and are not directly applied in downlink [179].

7.5.3   Edge Cloud

When discussing cloud RAN, the related topic of shifting services away from the network core to the network edge arises. *Edge cloud* is about bringing the cloud to the RAN, rather than shifting RAN to a big, centralized cloud. It is a distributed cloud computing platform at the base stations located in the cell sites that can help accelerate content delivery or can be used for management purposes. Intel is seeking to create an edge cloud of IP processing power to handle network tasks and data applications. By pushing processing and intelligence from gateways or core networks into the base station itself [186]. The main advantage of the edge cloud is the fact that it does not require investment in high-performance fiber backhaul or huge servers. Services running on RAN enable network operators to optimize backhaul network utilization, thus improving the user QoE.

It is commonly agreed that to deliver a strong mobile quality of experience it is required to bring the signal closer to the user to handle increasing amount of signaling due to the web and new applications traffic, as well as prioritizing certain transactions and data types. Then, the question that arises is, while the access points get closer to the UEs, where should the big amount of data processing and network tasks go on? Should we put it, too, as close to the end user as possible, or centralize in huge core network servers, and even virtualized in the cloud?

The answer depends mainly on the business model and most likely will be a combination of techniques in different locations or services. Depending on where their key strengths lie, different vendors have different positions. RAN suppliers whose core network offerings have been under pressure prefer to push edge cloud, while Cisco and enterprise integrators like IBM like centralization and virtualization and Intel take both sides [187, 188]. In general, it is expected large networks need both approaches, i.e., distributed or centralized intelligence.

7.6   Cloud RAN Market

Intel and China Mobile has been working on the development of C-RAN since 2009 [8, 42]. Today, many vendors and operators are carrying research and developing an architecture for C-RAN. Figure 7.7 illustrates C-RAN R&D activity levels for several leading vendors and operators. At mobile world congress 2014 (MWC'14),

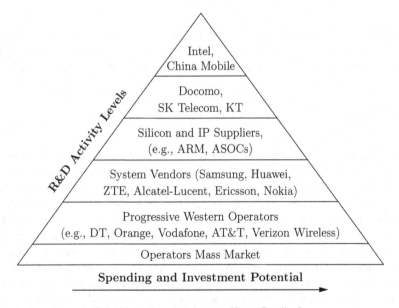

Fig. 7.7 Key players in C-RAN development [source: Heavy Reading]

China Mobile and Alcatel-Lucent demonstrated an NFV-based LTE BBU. Holding major trials across the country, China Mobile is expected to incorporate C-RAN in its commercially deployed networks between 2015 and 2016. China Mobile's first commercial deployment of a C-RAN for TD-LTE services use Ericsson's fiber fronthaul solution [189].

In South Korea, SK Telecom and KT, have implemented centralized base station architectures for their LTE networks. SK Telecom is working on vRAN technology to innovate the structure of the next-generation base station. In October 2013, SK Telecom announced that it has signed a memorandum of understanding with Intel Korea to develop virtualized RAN (vRAN). It loads various functions of a network in a virtualized software on universal hardware, e.g., computer CPUs. Japanese operator NTT Docomo joined C-RAN enthusiasts in 2013 by announcing the adoption of the structure for its LTE-A rollout.

Today all major vendors have their own CRAN solutions. MWC'16 commenced with a buzz of excitement surrounding Nokia, announcing AirScale RAN which can use any architecture topology (distributed RAN, centralized RAN or cloud RAN). Also, in February 2016, in a live demonstration of cloud RAN, Ericsson introduced its cloud RAN solution which offers distributed, centralized, elastic, and virtualized RAN.

Today, one main challenge for global adoption of centralized RAN is fronthaul requirements, and in the years to come, we expect to see more developments related to wireless fronthaul [141, 190]. Transferring huge volume of data from distributed RRHs to a central processing room (BBU room) requires extremely high bandwidth and low-latency fiber links which have not yet been deployed but are under development in most regions outside East Asia. Not having major fronthaul barriers, C-RAN presence is expected to continue to increase across the Asian region.

7.7 Summary

Cloud RAN centralizes RAN functionality into a shared resource pool or cloud and is seen as one of the enablers of 5G mobile networks. In this chapter, it is argued that C-RAN can offer a wide range of potentials and benefits and a compelling value proposition to operators by consolidating individual base station processing into a single or regional server. Compared to the conventional RAN, in C-RAN resource scaling is easier, resource utilization is higher, and energy consumption is less. Thus, C-RAN implementation scenarios provides a clear path to CAPEX/OPEX reduction. Also, the challenges and research frontiers of this rapidly growing field are reviewed and current market status is discussed in this chapter.

References

1. Ericsson mobility report (2016), https://www.ericsson.com/res/docs/2016/ericsson-mobility-report-2016.pdf. Accessed Nov 2016
2. A. De Domenico, E. Calvanese Strinati, A. Capone, Enabling green cellular networks: a survey and outlook. Comput. Commun. **37**, 5–24 (2014)
3. Network functions virtualization – everything old is new again (2016), http://www.f5.com/pdf/white-papers/service-provider-nfv-white-paper.pdf. Accessed Nov 2016
4. Network functions virtualization – challenges and solutions (2016), http://www.tmcnet.com/tmc/whitepapers/documents/whitepapers/2013/9377-network-functions-virtualization-challenges-solutions.pdf. Accessed Nov 2016
5. I. Grigorik, *High Performance Browser Networking: What Every Web Developer Should Know About Networking and Web Performance* (O'Reilly Media Inc, Newton, 2013)
6. Z. Ghebretensaé, J. Harmatos, K. Gustafsson, Mobile broadband backhaul network migration from TDM to carrier Ethernet. IEEE Commun. Mag. **48**(10), 102–109 (2010)
7. Alcatel-Lucent, A new era of mobile backhaul (2016), https://resources.alcatel-lucent.com/asset/163517. Accessed Oct 2016
8. C-RAN: the road towards green RAN (2011), http://labs.chinamobile.com/cran/wp-content/uploads/2014/06/20140613-C-RAN-WP-3.0.pdf. Accessed Nov 2016
9. E. Dahlman, S. Parkvall, J. Skold, *4G: LTE/LTE-Advanced for Mobile Broadband* (Academic Press, New York, 2014)
10. Global mobile network traffic–a summary of recent trends, GSMA Documents (2011), http://www.gsma.com/spectrum/wp-content/uploads/2012/03/analysysmasonpaperonglobalmobilenetworktrafficcorectedjuly11.pdf. Accessed Nov 2016
11. Ericsson mobility report (2012), http://www.ericsson.com/res/docs/2012/ericsson-mobility-report-november-2012.pdf. Accessed Nov 2016
12. Ericsson mobility report (2014), http://www.ericsson.com/res/docs/2014/ericsson-mobility-report-june-2014.pdf. Accessed Nov 2016
13. The zettabyte era-trends and analysis (2014), http://www.cisco.com/c/en/us/solutions/collateral/service-provider/visual-networking-index-vni/VNI_Hyperconnectivity_WP.html. Accessed Nov 2016
14. Cisco visual networking index: Forecast and methodology, 2013–2018 (2014), http://www.cisco.com/c/en/us/solutions/collateral/service-provider/ip-ngn-ip-next-generation-network/white_paper_c11-481360.pdf. Accessed Nov 2016
15. K. Sundaresan, M.Y. Arslan, S. Singh, S. Rangarajan, S.V. Krishnamurthy, Fluidnet: a flexible cloud-based radio access network for small cells, in *Proceedings of the 19th Annual International Conference on Mobile Computing & Networking* (ACM, 2013), pp. 99–110

© Springer International Publishing AG 2017
M. Vaezi and Y. Zhang, *Cloud Mobile Networks*, Wireless Networks,
DOI 10.1007/978-3-319-54496-0

16. Alcatel-Lucent, A new RAN for a new era of mobility, http://www2.alcatel-lucent.com/knowledge-center/public_files/waa/Wireless-All-Around-New-RAN-Era.pdf.   Accessed Nov 2016
17. OpEx to overtake CapEx by 2015 U.S. LTE forecast, http://www.aglmediagroup.com/opex-to-overtake-capex-by-2015-u-s-lte-forecast/. Accessed July 2016
18. M. Rahman, C. Despins, S. Affes, Analysis of capex and opex benefits of wireless access virtualization, in *Proceedings of IEEE International Conference on Communications Workshops (ICC)* (IEEE, 2013), pp. 436–440
19. K. Johansson, A. Furuskar, P. Karlsson, J. Zander, Relation between base station characteristics and cost structure in cellular systems, in *15th IEEE International Symposium on Personal, Indoor and Mobile Radio Communications (PIMRC)*, vol. 4 (IEEE, 2004), pp. 2627–2631
20. Kim, The economics of the thousand times challenge: Spectrum, efficiency and small cells (2012), http://techneconomyblog.com/tag/hspa/. Accessed July 2016
21. Small cell market status, Informa Telecoms and Media (2012)
22. R. Bendlin, T. Ekpenyong, D. Greenstreet, Paving the path for wireless capacity expansion, Texas Instruments, http://www.tmcnet.com/tmc/whitepapers/documents/whitepapers/2013/6924-alcatel-lucent-metro-cells-bigger-picture.pdf. Accessed July 2016
23. J.G. Andrews, Seven ways that HetNets are a cellular paradigm shift. IEEE Commun. Mag. **51**(3), 136–144 (2013)
24. J.G. Andrews, H. Claussen, M. Dohler, S. Rangan, M.C. Reed, Femtocells: past, present, and future. IEEE J. Sel. Areas Commun. **30**(3), 497–508 (2012)
25. M.P. Mills, The cloud begins with coal: Big data, big networks, big infrastructure, and big power, http://www.tech-pundit.com/wp-content/uploads/2013/07/Cloud_Begins_With_Coal.pdf?c761ac. Accessed July 2016
26. Bracing for the cloud, http://thebreakthrough.org/index.php/programs/economic-growth/bracing-for-the-cloud. Accessed July 2016
27. More data, less energy: Making network standby more efficient in billions of connected devices, The International Energy Agency (IEA) report (2014), http://www.iea.org/publications/freepublications/publication/moredata_lessenergy.pdf. Accessed July 2016
28. M.A. Marsan, L. Chiaraviglio, D. Ciullo, M. Meo, Optimal energy savings in cellular access networks, in *Proc IEEE International Conference on Communications (ICC) Workshops* (2009), pp. 1–5
29. V.H. Ward, L. Sofie, L. Bart, C. Didier, P. Mario, D. Piet, Trends in worldwide ict electricity consumption from 2007 to 2012. Comput. Commun. **50**(9), 64–76 (2014)
30. E. Oh, B. Krishnamachari, X. Liu, Z. Niu, Toward dynamic energy-efficient operation of cellular network infrastructure. IEEE Commun. Mag. **49**(6), 56–61 (2011)
31. G. Peter, F. Simon, Mobile basestations: reducing energy, in *Engineering and Technology Magazine*, vol. 6 (2011)
32. R. Rubenstein, Green gauges, Technology Trends (2011), pp. 12–14, http://content.yudu.com/A1rcux/TTP0311/resources/13.htm. Accessed July 2016
33. S. Sezer, S. Scott-Hayward, P.-K. Chouhan, B. Fraser, D. Lake, J. Finnegan, N. Viljoen, M. Miller, N. Rao, Are we ready for SDN? Implementation challenges for software-defined networks. IEEE Commun. Mag. **51**(7) (2013)
34. Vmware workstation, http://www.vmware.com/ca/en/products/workstation. Accessed May 2016
35. A.T. Campbell, H.G. De Meer, M.E. Kounavis, K. Miki, J.B. Vicente, D. Villela, A survey of programmable networks. ACM SIGCOMM Comput. Commun. Rev. **29**(2), 7–23 (1999)
36. D. Chisnall, *The Definitive Guide to the Xen Hypervisor* (Pearson Education, Boston, 2007)
37. D. de Oliveira, F.A. Baião, M. Mattoso, Towards a taxonomy for cloud computing from an e-science perspective, in *Cloud Computing*, pp. 47–62 (Springer, Berlin, 2010)
38. N. Antonopoulos, L. Gillam, *Cloud Computing* (Springer, Berlin, 2010)
39. A. Wang, M. Iyer, R. Dutta, G.N. Rouskas, I. Baldine, Network virtualization: technologies, perspectives, and frontiers. J. Lightwave Technol. **31**(4), 523–537 (2013)

40. N.M.K. Chowdhury, R. Boutaba, Network virtualization: state of the art and research challenges. IEEE Commun. Mag. **47**(7), 20–26 (2009)
41. N. Chowdhury, R. Boutaba, A survey of network virtualization. Comput. Netw. **54**(5), 862–876 (2010)
42. A.P. Bianzino, C. Chaudet, D. Rossi, J. Rougier, A survey of green networking research. IEEE Commun. Surv. Tutor. **14**(1), 3–20 (2012)
43. Network virtualization–path isolation design guide, http://www.cisco.com/c/en/us/td/docs/solutions/Enterprise/Network_Virtualization/PathIsol.html. Accessed May 2016
44. Framework of network virtualization for future networks
45. S.A. Baset, H. Schulzrinne, An analysis of the skype peer-to-peer internet telephony protocol (2004), arXiv preprint arXiv:cs/0412017
46. E.K. Lua, J. Crowcroft, M. Pias, R. Sharma, S. Lim et al., A survey and comparison of peer-to-peer overlay network schemes. IEEE Commun. Surv. Tutor. **7**(1–4), 72–93 (2005)
47. J. Tyson, How virtual private networks work, http://www.communicat.com.au/wp-content/uploads/2013/04/how_vpn_work.pdf. Accessed May 2016
48. Tunneling or port forwarding, http://searchenterprisewan.techtarget.com/definition/tunneling. Accessed July 2016
49. E. Geier, How (and why) to set up a VPN today, http://www.pcworld.com/article/2030763/how-and-why-to-set-up-a-vpn-today.html. Accessed May 2016
50. P. Knight, C. Lewis, Layer 2 and 3 virtual private networks: taxonomy, technology, and standardization efforts. IEEE Commun. Mag. **42**(6), 124–131 (2004)
51. T. Takeda, Framework and requirements for layer 1 virtual private networks (2007)
52. Alcatel-Lucent, VPN Services: Layer 2 or Layer 3?
53. Network functions virtualisation - introductory white paper, in SDN and OpenFlow World Congress (2013), http://portal.etsi.org/nfv/nfv_white_paper2.pdf. Accessed May 2016
54. Network functions virtualisation, http://www.etsi.org/technologies-clusters/technologies/689-network-functions-virtualisation. Accessed May 2016
55. X. Wang, P. Krishnamurthy, D. Tipper, Wireless network virtualization, in *International Conference on Computing, Networking and Communications (ICNC)* (IEEE, 2013), pp. 818–822
56. H. Wen, P.K. Tiwary, T. Le-Ngoc, *Wireless Virtualization* (Springer, Berlin, 2013)
57. L. Doyle, J. Kibiłda, T.K. Forde, L. DaSilva, Spectrum without bounds, networks without borders. Proc. IEEE **102**(3), 351–365 (2014)
58. R. Chandra, P. Bahl, MultiNet: connecting to multiple IEEE 802.11 networks using a single wireless card. Proc. IEEE INFOCOM **2**, 882–893 (2004)
59. Microsoft Research, Connecting to multiple IEEE 802.11 networks with one WiFi card, http://research.microsoft.com/en-us/um/redmond/projects/virtualwifi/. Accessed August 2016
60. L. Xia, S. Kumar, X. Yang, P. Gopalakrishnan, Y. Liu, S. Schoenberg, X. Guo, Virtual WiFi: bring virtualization from wired to wireless, in *ACM SIGPLAN Notices*, vol. 46 (ACM, 2011), pp. 181–192
61. Y. Al-Hazmi, H. de Meer, Virtualization of 802.11 interfaces for wireless mesh networks, in *Proceedings of the 8th International Conference on Wireless On-demand Network Systems and Services (WONS)* (2011), pp. 44–51
62. F. Boccardi, O. Aydin, U. Doetsch, T. Fahldieck, H. Mayer, User-centric architectures: enabling CoMP via hardware virtualization, in *Proceedings of IEEE International Symposium on Personal Indoor and Mobile Radio Communications (PIMRC)* (2012), pp. 191–196
63. D. Lee, H. Seo, B. Clerckx, E. Hardouin, D. Mazzarese, S. Nagata, K. Sayana, Coordinated multipoint transmission and reception in LTE-advanced: deployment scenarios and operational challenges. IEEE Commun. Mag. **50**(2), 148–155 (2012)
64. G. Smith, A. Chaturvedi, A. Mishra, S. Banerjee, Wireless virtualization on commodity 802.11 hardware, in *Proceedings of the Second ACM International Workshop on Wireless Network Testbeds, Experimental Evaluation and Characterization* (ACM, 2007), pp. 75–82
65. A. Gudipati, D. Perry, L.E. Li, S. Katti, SoftRAN: software defined radio access network, in *Proceedings of the Second ACM SIGCOMM Workshop on Hot Topics in Software Defined Networking* (ACM, 2013), pp. 25–30

66. S. Katti, L.E. Li, *RadioVisor: a slicing plane for radio access networks, in Presented as part of the Open Networking Summit 2014 (ONS 2014)* (USENIX, Santa Clara, CA, 2014)
67. K. Nakauchi, K. Ishizu, H. Murakami, A. Nakao, H. Harada, Amphibia: a cognitive virtualization platform for end-to-end slicing, in *Proceedings of IEEE International Conference on Communications (ICC)* (IEEE, 2011), pp. 1–5
68. Z. Zhu, P. Gupta, Q. Wang, S. Kalyanaraman, Y. Lin, H. Franke, S. Sarangi, Virtual base station pool: towards a wireless network cloud for radio access networks, in *Proceedings of the 8th ACM International Conference on Computing Frontiers* (ACM, 2011), p. 34
69. J.G. Andrews, S. Buzzi, W. Choi, S. Hanly, A. Lozano, A.C. Soong, J. C. Zhang, What will 5g be? (2014), arXiv preprint arXiv:1405.2957
70. S. Singhal, G. Hadjichristofi, I. Seskar, D. Raychaudhri, Evaluation of UML based wireless network virtualization, in *Proceedings of the Next Generation Internet Networks* (2008)
71. M. Pearce, S. Zeadally, R. Hunt, Virtualization: Issues, security threats, and solutions. ACM Comput. Surv. **45**(2), 17 (2013)
72. M. Armbrust, A. Fox, R. Griffith, A.D. Joseph, R. Katz, A. Konwinski, G. Lee, D. Patterson, A. Rabkin, I. Stoica et al., A view of cloud computing. Commun. ACM **53**(4), 50–58 (2010)
73. B. Kees, M. Jeroen, P. Martin, *Testing Cloud Services: How to Test SaaS, PaaS & IaaS* (Rocky Nook Inc, 2013)
74. Amazon elastic compute cloud (amazon ec2), http://aws.amazon.com/ec2/. Accessed May 2016
75. R. Buyya, C. Vecchiola, S.T. Selvi, *Mastering Cloud Computing: Foundations and Applications Programming* (Morgan Kaufmann, San Francisco, 2013)
76. K. Ren, C. Wang, Q. Wang et al., Security challenges for the public cloud. IEEE Internet Comput. **16**(1), 69–73 (2012)
77. D. Zissis, D. Lekkas, Addressing cloud computing security issues. Future Gener. Comput. Syst. **28**(3), 583–592 (2012)
78. S. Angeles, Virtualization vs. Cloud Computing: What's the Difference?, http://www.businessnewsdaily.com/5791-virtualization-vs-cloud-computing.html. Accessed May 2016
79. Virtualization and cloud computing: Steps in the evolution from virtualization to private cloud infrastructure as a service, http://www.intel.com/content/dam/www/public/us/en/documents/guides/cloud-computing-virtualization-building-private-iaas-guide.pdf. Accessed June 2016
80. M. Forzati, C. Larsen, C. Mattsson, Open access networks, the Swedish experience, in *Proceedings of 12th IEEE International Conference on Transparent Optical Networks* (2010), pp. 1–4
81. R. Sherwood, G. Gibb, K. Yap, G. Appenzeller, M. Casado, N. McKeown, G. Parulkar, *Flowvisor: A network virtualization layer* (OpenFlow Switch Consortium, Technical Report, 2009)
82. M. Casado, T. Koponen, R. Ramanathan, S. Shenker, Virtualizing the network forwarding plane, in *Proceedings of the Workshop on Programmable Routers for Extensible Services of Tomorrow* (ACM, 2010)
83. M. Gerola, Enabling Network Virtualization in OpenFlow Networks through Virtual Topologies Generalization
84. http://www.openflow.org/
85. http://tools.ietf.org/html/draft-leymann-mpls-seamless-mpls-03
86. E.N. ISG, Network Functions Virtualisation (NFV) Architectural Framework, ETSI GS NFV 002 V1.1.1 (2013)
87. Y. Zhang, N. Beheshti, L. Beliveau, G. Lefebvre, R. Manghirmalani, R. Mishra, R. Patney, M. Shirazipour, R. Subrahmaniam, C. Truchan et al., Steering: a software-defined networking for inline service chaining, in *Proceedings of IEEE International Conference on Network Protocols* (2013), pp. 1–10
88. Z.A. Qazi, C.-C. Tu, L. Chiang, R. Miao, V. Sekar, M. Yu, Simple-fying middlebox policy enforcement using sdn, in *Proceedings of the ACM SIGCOMM 2013 conference on SIGCOMM* (ACM, 2013), pp. 27–38

89. L.E. Li, Z.M. Mao, J. Rexford, Toward software-defined cellular networks, in *Proceedings of the European Workshop on Software Defined Networking. EWSDN'12* (2012)
90. X. Jin, L. E. Li, L. Vanbever, J. Rexford, Softcell: scalable and flexible cellular core network architecture, in *Proceedings of the ACM Conference on Emerging Networking Experiments and Technologies, CoNEXT '13* (2013)
91. Y. Zhang, N. Beheshti, M. Tatipamula, On resilience of split-architecture networks, in *Proceedings of IEEE GLOBECOM 2011 - Next Generation Networking Symposium* (2011)
92. Network Functions Virtualisation, http://www.etsi.org/technologies-clusters/technologies/nfv. Accessed Nov 2016
93. Z.A. Qazi, C.-C. Tu, L. Chiang, R. Miao, V. Sekar, M. Yu, Simple-fying middlebox policy enforcement using sdn, in *Proceedings of the ACM SIGCOMM, SIGCOMM '13* (2013), pp. 27–38
94. S.K. Fayazbakhsh, L. Chiang, V. Sekar, M. Yu, J.C. Mogul, Enforcing network-wide policies in the presence of dynamic middlebox actions using flowtags, in *Proceedings of the USENIX Conference on Networked Systems Design and Implementation* (2014), pp. 533–546
95. Y. Zhang, N. Beheshti, L. Beliveau, G. Lefebvre, R. Manghirmalani, R. Mishra, R. Patneyt, M. Shirazipour, R. Subramaniam, C. Truchan, M. Tatipamula, Steering: a software-defined networking for inline service chaining, in *Proceedings of IEEE International Conference on Network Protocols* (2013), pp. 1–10
96. H. Moens, F. De Turck, VNF-P: a model for efficient placement of virtualized network functions, in *Proceedings of International Conference on Network and Service Management (CNSM)* (2014), pp. 418–423
97. M.C. Luizelli, L.R. Bays, L.S. Buriol, M.P. Barcellos, L.P. Gaspary, Piecing together the nfv provisioning puzzle: efficient placement and chaining of virtual network functions, in *Proceedings of IFIP/IEEE International Symposium on Integrated Network Management* (2015), pp. 98–106
98. S. Mehraghdam, M. Keller, H. Karl, Specifying and placing chains of virtual network functions, in *Proceedings of IEEE 3rd International Conference on Cloud Networking (CloudNet)* (2014), pp. 7–13
99. M.F. Bari, S.R. Chowdhury, R. Ahmed, R. Boutaba, On orchestrating virtual network functions, in *Proceedings of IEEE 11th International Conference on Network and Service Management* (2015), pp. 50–56
100. 3GPP, LTE; Feasibility study for Further Advancements for E-UTRA (LTE-Advanced). 3GPP TR 36.912 version 9.0.0 Release 9 (2009), http://www.etsi.org/deliver/etsi_tr/136900_136999/136912/09.00.00_60/tr_136912v090000p.pdf. Accessed Oct 2016
101. M. Laner, P. Svoboda, P. Romirer-Maierhofer, N. Nikaein, F. Ricciato, M. Rupp, A comparison between one-way delays in operating HSPA and LTE networks, in *Proceedings of 10th IEEE International Symposium on Modeling and Optimization in Mobile, Ad Hoc and Wireless Networks (WiOpt)* (2012), pp. 286–292
102. P. Chanclou, A. Pizzinat, F. Le Clech, T.-L. Reedeker, Y. Lagadec, F. Saliou, B. Le Guyader, L. Guillo, Q. Deniel, S. Gosselin et al., Optical fiber solution for mobile fronthaul to achieve cloud radio access network, in *IEEE Future Network & Mobile Summit* (2013), pp. 1–11
103. Alcatel-Lucent CDMA Distributed Base Station Portfolio, http://www3.alcatel-lucent.com/wps/DocumentStreamerServlet?LMSG_CABINET=Docs_and_Resource_Ctr&LMSG_CONTENT_FILE=Brochures/Q209_CDMADistribBSPortfolio_bro.pdf. Accessed Apr 2016
104. Active antenna system: Utilizing the full potential of radio sources in the spatial domain, www.huawei.com/ilink/en/download/HW_197969. Accessed June 2016
105. O. Tipmongkolsilp, S. Zaghloul, A. Jukan, The evolution of cellular backhaul technologies: current issues and future trends. IEEE Commun. Surv. Tutor. **13**(1), 97–113 (2011)
106. P. Briggs, R. Chundury, J. Olsson, Carrier ethernet for mobile backhaul. IEEE Commun. Mag. **48**(10), 94–100 (2010)
107. U. Kohn, Is fronthaul the future of mobile backhaul networks?, http://blog.advaoptical.com/is-fronthaul-the-future-of-mobile-backhaul-networks. Accessed Dec 2016

108. 3GPP TR 36.93, http://www.3gpp.org/DynaReport/36932.htm. Accessed Aug 2016
109. S. Khan, J. Edstam, B. Varga, J. Rosenberg, J. Volkering, M. Stumpert, The benefits of self-organizing backhaul networks, in *Ericsson Review* (2013)
110. R. Sue, A tale of two SONs: Unraveling D-SON and C-SON (2014), http://www.pipelinepub.com/cloud_and_network_virtualization/SON_D-Son_and_C-Son. Accessed Aug 2016
111. S. Feng, E. Seidel, Self-organizing networks (SON) in 3GPP long term evolution (2008), http://www.3g4g.co.uk/Lte/LTE_SON_WP_0805_Nomor.pdf. Accessed Aug 2016
112. CPRI Specification V6.0, http://www.cpri.info/downloads/CPRI_v_6_0_2013-08-30.pdf. Accessed Sept 2016
113. Open base station architecture initiative: BTS system reference document version 2.0, http://www.obsai.com/specs/OBSAI_System_Spec_V2.0.pdf. Accessed Sept 2016
114. K.F. Nieman, B.L. Evans, Time-domain compression of complex-baseband LTE signals for cloud radio access networks, in *Proceedings of IEEE Global Conference on Signal and Information Processing (GlobalSIP)* (2013), pp. 1198–1201
115. N. Bhushan, J. Li, D. Malladi, R. Gilmore, D. Brenner, A. Damnjanovic, R. Sukhavasi, C. Patel, S. Geirhofer, Network densification: the dominant theme for wireless evolution into 5G. IEEE Commun. Mag. **52**, 82–89 (2014)
116. J. Hoydis, S. Ten Brink, M. Debbah, Massive MIMO in the UL/DL of cellular networks: how many antennas do we need? IEEE J. Sel. Areas Commun. **31**(2), 160–171 (2013)
117. D. Tse, P. Viswanath, *Fundamentals of Wireless Communication* (Cambridge University Press, Cambridge, 2005)
118. D. Gesbert, M. Kountouris, R.W. Heath, C.-B. Chae, T. Salzer, Shifting the mimo paradigm. IEEE Signal Process. Mag. **24**(5), 36–46 (2007)
119. E.G. Larsson, O. Edfors, F. Tufvesson, T.L. Marzetta, Massive MIMO for next generation wireless systems. IEEE Commun. Mag. **52**(2), 186–195 (2014)
120. L. Lu, G.Y. Li, A.L. Swindlehurst, A. Ashikhmin, R. Zhang, An overview of massive MIMO: benefits and challenges. IEEE J. Sel. Top. Signal Process. **8**(5), 742–758 (2014)
121. F. Rusek, D. Persson, B.K. Lau, E.G. Larsson, T.L. Marzetta, O. Edfors, F. Tufvesson, Scaling up MIMO: opportunities and challenges with very large arrays. IEEE Signal Process. Mag. **30**(1), 40–60 (2013)
122. Y. Saito, Y. Kishiyama, A. Benjebbour, T. Nakamura, A. Li, K. Higuchi, Non-orthogonal multiple access (NOMA) for cellular future radio access, in *Proceedings of IEEE 77th Vehicular Technology Conference (VTC Spring)* (2013), pp. 1–5
123. 3GPP TD RP-150496, Study on Downlink Multiuser Superposition Transmission (2015)
124. H. Nikopour, H. Baligh, Sparse code multiple access, in *Proceedings of the IEEE International Symposium on Personal, Indoor, and Mobile Radio Communications (PIMRC)* (2013), pp. 332–336
125. W. Shin, M. Vaezi, J. Lee, H.V. Poor, On the number of users served in MIMO-NOMA cellular networks, in *Proceedings of IEEE International Symposium on Wireless Communication Systems (ISWCS)* (2016), pp. 638–642
126. L. Dai, B. Wang, Y. Yuan, S. Han, C.-L. I, Z. Wang, Non-orthogonal multiple access for 5G: solutions, challenges, opportunities, and future research trends. IEEE Commun. Mag. **53**(9), 74–81 (2015)
127. W. Shin, M. Vaezi, B. Lee, D.J. Love, J. Lee, H.V. Poor, Non-orthogonal multiple access in multi-cell networks: theory, performance, and practical challenges, IEEE Commun. Mag. (2017)
128. M. Vaezi, H.V. Poor, Simplified Han-Kobayashi region for one-sided and mixed Gaussian interference channels, in *Proceedings of IEEE International Conference on Communications (ICC)* (2016)
129. S. Rangan, T.S. Rappaport, E. Erkip, Millimeter wave cellular wireless networks: potentials and challenges (2014), arXiv preprint arXiv:1401.2560
130. Alcatel-Lucent, Metro cells the bigger picture (2012), http://www.tmcnet.com/tmc/whitepapers/documents/whitepapers/2013/6924-alcatel-lucent-metro-cells-bigger-picture.pdf. Accessed July 2016

131. D. Lopez-Perez, I. Guvenc, G. De La Roche, M. Kountouris, T.Q. Quek, J. Zhang, Enhanced intercell interference coordination challenges in heterogeneous networks. IEEE Wirel. Commun. **18**(3), 22–30 (2011)

132. Small cell market highlights, Small Cell Forum (2014), http://www.scf.io/en/documents/050_-_Market_status_statistics_Q1_2014_-Mobile_Experts.php. Accessed Nov 2016

133. Small cell market highlights, Small Cell Forum (2016), http://scf.io/en/documents/050_-_Market_status_report_Feb_2016_-_Mobile_Experts.php. Accessed Nov 2016

134. Cisco, Trends in wireless network densification, https://communities.cisco.com/community/solutions/sp/mobility/blog/2014/04/21/trends-in-wireless-network-densification. Accessed Oct 2016

135. J. Zhang, G. De la Roche et al., *Femtocells: Technologies and Deployment* (Wiley Online Library, London, 2010)

136. Huawei, LTE small cell v.s. WiFi user experience, www.huawei.com/ilink/en/download/HW_323974. Accessed Sept 2016

137. A. Brydon, 3GPP network sharing enhancements for LTE, http://www.unwiredinsight.com/2013/3gpp-lte-ran-sharing-enhancements. Accessed Mar 2016

138. Alcatel-Lucent, Network Sharing in LTE: Opportunity & Solutions (2010), http://images.tmcnet.com/online-communities/ngc/pdfs/Network-Sharing-in-LTE-Opportunity-and-solutions.pdf. Accessed Oct 2016

139. A. Khan, W. Kellerer, K. Kozu, M. Yabusaki, Network sharing in the next mobile network: TCO reduction, management flexibility, and operational independence. IEEE Commun. Mag. **49**(10), 134–142 (2011)

140. P. Rost, I. Berberana, A. Maeder, H. Paul, V. Suryaprakash, M. Valenti, D. Wübben, A. Dekorsy, G. Fettweis, Benefits and challenges of virtualization in 5G radio access networks. IEEE Commun. Mag. **53**(12), 75–82 (2015)

141. F. Rayal, Cloud RAN pros, cons and disruptive potential, http://the-mobile-network.com/2014/01/cloud-ran-pros-cons-and-disruptive-potential/. Accessed Sept 2016

142. D. Staehle, K. Leibnitz, P. Tran-Gia, Source traffic modeling of wireless applications

143. A. Klemm, C. Lindemann, M. Lohmann, Traffic modeling and characterization for umts networks. Proc. IEEE Glob. Telecommun. Conf. (GLOBECOM) **3**, 1741–1746 (2001)

144. M. Madhavan, P. Gupta, M. Chetlur, Quantifying multiplexing gains in a wireless network cloud, in *Proceedings of IEEE International Conference on Communications (ICC)* (2012), pp. 3212–3216

145. S. Bhaumik, S.P. Chandrabose, M.K. Jataprolu, G. Kumar, A. Muralidhar, P. Polakos, V. Srinivasan, T. Woo, CloudIQ: a framework for processing base stations in a data center, in *Proceedings of the 18th Annual International Conference on Mobile Computing and Networking* (ACM, 2012), pp. 125–136

146. T. Werthmann, H. Grob-Lipski, M. Proebster, Multiplexing gains achieved in pools of baseband computation units in 4G cellular networks, in *Proceedings of IEEE 24th International Symposium on Personal Indoor and Mobile Radio Communications (PIMRC)* (2013), pp. 3328–3333

147. A. Checko, H. Holm, H. Christiansen, Optimizing small cell deployment by the use of C-RANs, in *Proceedings of 20th European Wireless Conference* (2014), pp. 1–6

148. S. Namba, T. Matsunaka, T. Warabino, S. Kaneko, Y. Kishi, Colony-RAN architecture for future cellular network, in *Proceedings of IEEE Future Network & Mobile Summit (FutureNetw)* (2012), pp. 1–8

149. W. Yu, T. Kwon, C. Shin, Statistical multiplexing gain analysis of heterogeneous virtual base station pools in cloud radio access networks. IEEE Trans. Wirel. Commun. **15**(8), 5681–5694 (2016)

150. D. Gesbert, S. Hanly, H. Huang, S. Shamai Shitz, O. Simeone, W. Yu, Multi-cell MIMO cooperative networks: a new look at interference. IEEE J. Sel. Areas Commun. **28**(9), 1380–1408 (2010)

151. S. Sun, Q. Gao, Y. Peng, Y. Wang, L. Song, Interference management through CoMP in 3GPP LTE-advanced networks. IEEE Wirel. Commun. **20**(1), 59–66 (2013)

152. W. Shin, M. Vaezi, B. Lee, D. J. Love, J. Lee, V. Poor, Coordinated beamforming for multi-cell mimo-noma. IEEE Commun. Lett. **21**(1), 84–87 (2017)
153. E. Björnson, E. Jorswieck, Optimal resource allocation in coordinated multi-cell systems. Found. Trends Commun. Inf. Theory **9**(2–3) (2013)
154. Coordinated multi-point operation for LTE physical layer aspects, 3GPP TR 36.819 V11.2.0, http://www.3gpp.org/DynaReport/36819.htm. Accessed Sept 2016
155. W. Nam, D. Bai, J. Lee, I. Kang, Advanced interference management for 5G cellular networks. IEEE Commun. Mag. **52**(5), 52–60 (2014)
156. INFSO-ICT-317941 iJOIN, State-of-the-Art of and promising candidates for PHY layer approaches on access and backhaul network, http://www.ictijoin.eu/wp-content/uploads/2014/01/D2.1.pdf. Accessed Aug 2016
157. J. Wu, S. Rangan, H. Zhang, *Green Communications: Theoretical Fundamentals, Algorithms and Applications* (CRC Press, Boca Raton, 2012)
158. M. Yang, Y. Li, D. Jin, L. Su, S. Ma, L. Zeng, OpenRAN: a software-defined ran architecture via virtualization, in *Proceedings of the ACM SIGCOMM* (2013), pp. 549–550
159. O. Oyman, S. Singh, Quality of experience for http adaptive streaming services. IEEE Commun. Mag. **50**(4), 20–27 (2012)
160. Human Factors (HF); Quality of Experience (QoE) requirements for real-time communication services, ETSI TR 102 643 V1.0.1, http://www.etsi.org/deliver/etsi_tr/102600_102699/102643/01.00.01_60/tr_102643v010001p.pdf. Accessed Aug 2016
161. M. Nohrborg, LTE Overview, 3GPP, http://www.3gpp.org/technologies/keywords-acronyms/98-lte. Accessed Aug 2016
162. A. Larmo, M. Lindstrom, M. Meyer, G. Pelletier, J. Torsner, H. Wiemann, The LTE link-layer design. IEEE Commun. Mag. **47**(4), 52–59 (2009)
163. J. Zyren, Overview of the 3GPP long term evolution physical layer
164. Distributed base stations, http://www.speed2design.com/reports/DBS_Application_Note.pdf. Accessed Apr 2016
165. C.F. Lanzani, G. Kardaras, D. Boppana, Remote radio heads and the evolution towards 4G networks
166. Radisys, Evaluating Cloud RAN Implementation Scenarios (2014), https://www.thinksmallcell.com/send/3-white-papers/26-evaluating-cloud-ran-implementation-strategies.html?option=com_jdownloads. Accessed Nov 2016
167. P. Rost, C.J. Bernardos, A. Domenico, M. Girolamo, M. Lalam, A. Maeder, D. Sabella et al., Cloud technologies for flexible 5G radio access networks. IEEE Commun. Mag. **52**(5), 68–76 (2014)
168. D. Sabella, P. Rost, Y. Sheng, E. Pateromichelakis, U. Salim, P. Guitton-Ouhamou, M. Di Girolamo, G. Giuliani, RAN as a service: challenges of designing a flexible RAN architecture in a cloud-based heterogeneous mobile network. Future Network and Mobile Summit (FutureNetworkSummit) (2013), pp. 1–8
169. C.J. Bernardos, A. De Domenico, J. Ortin, P. Rost, D. Wubben, Challenges of designing jointly the backhaul and radio access network in a cloud-based mobile network, in *Proceedings of IEEE Future Network and Mobile Summit (FutureNetworkSummit)* (2013), pp. 1–10
170. J. Madden, Cloud RAN or small cells?, http://www.fiercewireless.com/tech/story/madden-cloud-ran-or-small-cells/2013-04-30. Accessed Oct 2016
171. D. Samardzija, J. Pastalan, M. MacDonald, S. Walker, R. Valenzuela, Compressed transport of baseband signals in radio access networks. IEEE Trans. Wirel. Commun. **11**(9), 3216–3225 (2012)
172. Nokia, "LTE-capable transport: a quality user experience demands an end-to-end approach, 3GPP TR 36.912 version 9.0.0 Release 9, http://nsn.com/index.php?q=system/files/document/lte_transport_requirements.pdf. Accessed Oct 2016
173. S. Nanba, A. Agata, A new IQ data compression scheme for front-haul link in Centralized RAN, in *Proceedings of 24th IEEE International Symposium on Personal, Indoor and Mobile Radio Communications (PIMRC Workshops)* (2013), pp. 210–214

174. S. Shin, S.M. Shin, Why should jitter be minimized in CPRI fronthaul?, http://www.netmanias. com/en/?m=view&id=blog&no=6460. Accessed Oct 2016

175. Fronthaul compression for emerging C-RAN and small cell networks (2013), https://www. idt.com/document/whp/front-haul-compression-emerging-c-ran-and-small-cell-networks. Accessed Oct 2016

176. 3GPP, LTE; Evolved Universal Terrestrial Radio Access (E-UTRA); Base Station (BS) radio transmission and reception. 3GPP TS 36.104 version 11.9.0 Release 11 (2014), http:// www.etsi.org/deliver/etsi_ts/136100_136199/136104/11.09.00_60/ts_136104v110900p. pdf. Accessed Oct 2016

177. A. Vosoughi, M. Wu, J.R. Cavallaro, Baseband signal compression in wireless base stations, in *Proceedings of IEEE Global Communications Conference (GLOBECOM)* (2012), pp. 4505–4511

178. A. Del Coso, S. Simoens, Distributed compression for MIMO coordinated networks with a backhaul constraint. IEEE Trans. Wirel. Commun. **8**(9), 4698–4709 (2009)

179. M. Peng, Y. Sun, X. Li, Z. Mao, C. Wang, Recent advances in cloud radio access networks: system architectures, key techniques, and open issues. IEEE Commun. Surv. Tutor. **18**(3), 3–20 (2016)

180. B. Guo, W. Cao, A. Tao, D. Samardzija, LTE/LTE-a signal compression on the CPRI interface. Bell Labs Tech. J. **18**(2), 117–133 (2013)

181. S.-H. Park, O. Simeone, O. Sahin, S. Shamai, Robust and efficient distributed compression for cloud radio access networks. IEEE Trans. Veh. Technol. **62**(2), 692–703 (2013)

182. J. Lorca, L. Cucala, Lossless compression technique for the fronthaul of LTE/LTE-advanced cloud-RAN architectures, in *Proceedings of 14th IEEE International Symposium and Workshops on a World of Wireless, Mobile and Multimedia Networks (WoWMoM)* (2013), pp. 1–9

183. P.L. Dragotti, M. Gastpar, *Distributed Source Coding: Theory, Algorithms, and Applications* (Academic Press, New York, 2009)

184. Z. Xiong, A.D. Liveris, S. Cheng, Distributed source coding for sensor networks. IEEE Signal Process. Mag. **21**, 80–94 (2004)

185. M. Vaezi, F. Labeau, Distributed source-channel coding based on real-field BCH codes, IEEE Trans. Signal Process. **62**(5), 1171–1184 (2014)

186. C. Gabriel, Intel takes both sides in the 4G debate – Cloud-RAN or edge cloud? (2011), http://www.rethink-wireless.com/2011/5/25/Intel-takes-both-sides-in-the-4G-debateCloudRAN-or-edge-cloud-page2. Accessed Aug 2016

187. Intel Heterogeneous Network Solution Brief, http://www.intel.com/content/dam/www/ public/us/en/documents/solution-briefs/communications-heterogeneous-network-brief.pdf. Accessed Oct 2016

188. Increasing Network ROI with Cloud Computing at the Edge, http://www.intel.com/content/ dam/www/public/us/en/documents/solution-briefs/intelligent-base-stations-increase-cloud-computing-roi-brief.pdf. Accessed October 2016

189. Ericsson, Ericsson's fiber fronthaul solution deployed for China Mobile's LTE C-RAN, http://www.ericsson.com/news/140707-ericssons-fiber-fronthaul-solution-deployed_244099436_c. Accessed Oct 2016

190. Y. Lin, L. Shao, Z. Zhu, Q. Wang, R.K. Sabhikhi, Wireless network cloud: architecture and system requirements. IBM J. Res. Dev. **54**(1), 1–4 (2010)

Printed in the United States
By Bookmasters